Quality Assurance in Distance Education and E-learning

Thank you for choosing a SAGE product! If you have any comment, observation or feedback, I would like to personally hear from you. Please write to me at contactceo@sagepub.in

—Vivek Mehra, Managing Director and CEO,
SAGE Publications India Pvt Ltd, New Delhi

Bulk Sales

SAGE India offers special discounts for purchase of books in bulk. We also make available special imprints and excerpts from our books on demand.

For orders and enquiries, write to us at

Marketing Department
SAGE Publications India Pvt Ltd
B1/I-1, Mohan Cooperative Industrial Area
Mathura Road, Post Bag 7
New Delhi 110044, India
E-mail us at marketing@sagepub.in

Get to know more about SAGE, be invited to SAGE events, get on our mailing list. Write today to marketing@sagepub.in

This book is also available as an e-book.

Quality Assurance in Distance Education and E-learning

Challenges and Solutions from Asia

Edited by
Insung Jung, Tat Meng Wong, and Tian Belawati

International Development Research Centre
Ottawa • Cairo • Montevideo • Nairobi • New Delhi

⑨SAGE www.sagepublications.com
Los Angeles • London • New Delhi • Singapore • Washington DC

First published in 2013 by

SAGE Publications India Pvt Ltd
B1/I-1 Mohan Cooperative Industrial Area
Mathura Road, New Delhi 110 044, India
www.sagepub.in

SAGE Publications Inc
2455 Teller Road
Thousand Oaks, California 91320, USA

SAGE Publications Ltd
1 Oliver's Yard, 55 City Road
London EC1Y 1SP, United Kingdom

SAGE Publications Asia-Pacific Pte Ltd
33 Pekin Street
#02-01 Far East Square
Singapore 048763

✳ IDRC | CRDI

International
Development Research
Centre
P.O. Box 8500
Ottawa, ON
Canada K1G 3H9
www.idrc.ca
info@idrc.ca
ISBN (e-book)
987-81-321-1006-4

Published by Vivek Mehra for SAGE Publications India Pvt Ltd, Photo typeset in 10/12 Palatino by Tantla Composition Pvt Ltd, Chandigarh and printed at De-Unique, New Delhi.

Library of Congress Cataloging-in-Publication Data
Quality assurance in distance education and E-learning : challenges and solutions from Asia / edited by Insung Jung, Tat Meng Wong, and Tian Belawati.
    pages cm.
Includes bibliographical references and index.
    1. Distance education—Asia   2. Distance education—Asia—Computer-assisted instruction.   3. Distance education—Standards.   4. Quality assurance.   I. Jung, Insung, 1959–   II. Jung, Insung.

LC5808.A78J86   371.35095—dc23   2012   2012045160

ISBN: 978-81-321-1006-4 (HB)

The SAGE Team: Rudra Narayan, Rohini Rangachari Karnik, Nand Kumar Jha and Dally Verghese

Contents

List of Tables and Figures   vii
List of Abbreviations   ix
Foreword by Asha Kanwar   xvii
Preface   xxi

Part 1: *A Systems or Balanced Approach to Quality Assurance*

1. Singapore's SIM University
   Cheong Hee Kiat   3
2. Thailand's Sukhothai Thammathirat Open University
   Pranee Sungkatavat and Theppasak Boonyarataphan   25
3. Open University of Hong Kong
   Robert Edward Butcher   42
4. S. Korea's Hanyang Cyber University
   Yeonwook Im   57

Part 2: *Ensuring the Quality of Management Processes*

5. Indonesia's Universitas Terbuka
   Sri Y.P.K. Hardini, Deetje Sunarsih, Any Meilani, and Tian Belawati   81
6. China's Peking University School of Distance Learning for Medical Education
   Chen Li, Shen Xinyi, Gao Shuping, and Liu Yiguang   94
7. Mongolian e-Knowledge
   Sanjaa Baigaltugs   105
8. S. Korea's AutoEver
   Hae-Deok Song and Cheolil Lim   118

Part 3: *Focusing on Instructional Design and Pedagogy*

9. Japan's Kumamoto University Online Graduate School
   Katsuaki Suzuki   139

10. Open University of China
    Li Yawan, Yang Tingting, and Niu Ben   155
11. India's Indira Gandhi National Open University
    Pema Eden Samdup and Rose Nembiakkim   169
12. University of the Philippines Open University
    Patricia B. Arinto   182

Part 4: *Assuring Quality of Learning Support and Assessment*

13. Malaysia's Wawasan Open University
    Tat Meng Wong and Teik Kooi Liew   199
14. Virtual University of Pakistan
    Naveed Akhtar Malik   220

Part 5: *Outcomes and Performance Measurement*

15. Open University of Sri Lanka
    Uma Coomaraswamy   241
16. Open University Malaysia
    Anuwar Ali and Mansor Bin Fadzil   258

Concluding Remarks: Future Policy Directions
Insung Jung   275

About the Editors and Contributors   289
Index   298

List of Tables and Figures

TABLES

2.1   Nine components and indicators for internal QA for HEIs   26

4.1   Rate of annual enrollment   60
4.2   Evaluation criteria for qualitative analysis   62
4.3   Educational system model of cyber universities   63
4.4   Monitoring criteria and scoring method   64
4.5   HYCU's undergraduate programs   66
4.6   HYCU's graduate programs   66

5.1   QA components and number of QA manuals at UT   87
5.2   List of ISO 9001:2008 certificates   89

6.1   QA duties of offices and departments   99

8.1   QA tools and their uses during specific stages of
      program development   126
8.2   QA tools and their uses during specific
      stages of program implementation   128
8.3   Learner support tools in LMS   131

9.1   Results of QAAS by agency   141
9.2   Sample decisions made during Course Team meetings   146
9.3   GSIS course design policies   147

13.1   Nine areas of evaluation for program
       accreditation and institutional audit   201

15.1   OUSL's experience with the QA toolkit   251

FIGURES

1.1   UniSIM organization structure   7
1.2   UniSIM's academic quality assurance framework   10

2.1   Relationship between internal and external
       QA measures at STOU   29
2.2   Integration of a total quality management
       model in STOU's QA system   39

4.1   Cyber universities' student distribution by age   59
4.2   Cyber universities' student distribution by
       academic background   59
4.3   Initial institutional accreditation procedure   61
4.4   How to take a course at HYCU   67
4.5   Process of content development   71

8.1   QA organizational structure for the AE E-learning Unit   121
8.2   E-learning course evaluation process   123
8.3   Individualized e-learning cycle in an LMS environment   124
8.4   E-learning evaluation process for unexpected demands   130

9.1   Overall QA process for GSIS   144

11.1   Program approval process at IGNOU   174
11.2   Stages involved in course development   175
11.3   Academic bodies overlooking QA at IGNOU   178

12.1   Flow chart for academic proposals   186

13.1   WOU's multi-entry and multi-exit progression pathways   203
13.2   QA intervention during program planning
        and development   209
13.3   QA intervention during the course development process   210
13.4   Tutor training and management system   211
13.5   QA in assignment management   213
13.6   QA in the management of examinations   214

14.1   Essay-type question definition   231
14.2   Multiple-choice type question definition   232
14.3   Exam entrance slip after creating datesheet   233
14.4   Setting question paper criteria   234
14.5   Grading an essay type question with a rubric
        provided by the question bank   235

List of Abbreviations

AAC	Academic Affairs Committee, University of the Philippines
AACCUP	Accrediting Association of Chartered Colleges and Universities of the Philippines, Philippine
AAOU	Asian Association of Open Universities
ADB	Asian Development Bank
ADB-DEMP	Asian Development Bank-Distance Education Modernization Project
AE	Hyundai AutoEver, S. Korea
AF	Associate Faculty, SIM University
AICTE	All India Council of Technical Education
APG	Advisory Peer Group
APQN	Asia Pacific Quality Network
ARC	Annual Reporting and Censorship, China
ASEAN	Association of Southeast Asian Nations
AU	Autonomous Universities
BAN-PT	Badan Akreditasi National Perguruan Tinggi, Indonesia or Indonesia's Natioral Accreditation Board for Higher Education
BSC	Balanced Scorecard
BSNP	Badan Standar Nasional Pendidikan, Indonesia
CA	Continuous Assessment
CC	Course Coordinator
CHED	Commission on Higher Education, Philippines
CIDT	Center for Instructional Design and Technology, Open University Malaysia
COL	Commonwealth of Learning
COL-RIM	Commonwealth of Learning Review and Implementation Model
COPIA	Code of Practice for Institutional Audit, Malaysia
COPPA	Code of Practice for Program Accreditation, Malaysia

CPAT	Course Production and Administration Team, STOU
CPC	Curriculum and Planning Committee, SIM University
CPE	Council of Private Education, SIM University
CRC	Camera Ready Copy, Indira Gandhi National Open University
CRM	Customer Relationship Management, SIM University
CSM	Center for Student Management, Open University Malaysia
CSMS	Computer Science and Management School, Mongolian University of Science and Technology
CT	Course Team
CU	Constituent Universities, University of the Philippines
CVCD	Committee of Vice-Chancellors and Directors, Sri Lanka
DE	Distance Education
DEC	Distance Education Council, India
DEIs	Distance Education Institutes
DEMP	Distance Education Modernization Project
DEP	Diploma Education Program, China's Peking University School of Distance Learning
DPP	Detailed Program Proposal, Open University of Hong Kong
DPP	Detailed Program Proposal, Malaysia's Wawasan Open University
DVC-A	Deputy Vice Chancellor (Academic), Malaysia's Wawasan Open University
EAS	Establishment-Approval System, Japan
ECA	External Course Assessor
EE	External Examiner, Open University of Hong Kong
EFQM	European Foundation for Quality Management
eLC	e-Learning Consortium
ETP	Education Technology and Publishing, Malaysia's Wawasan Open University
EUA	European University Association
FAAP	Federation of Accrediting Agencies of the Philippines, Philippine

FIC	Faculty-in-Charge, University of the Philippines-Open University
GBS	Goal-based Scenarios
GMC	Group Management Committee, Open University Malaysia
GSIS	Graduate School of Instructional Systems, Japan's Kumamoto University
HAU	Heads of Academic Units, Malaysia's Wawasan Open University
HEC	Higher Education Commission, Pakistan
HEFCE	Higher Education Funding Council of England
HEIs	Higher Education Institutions
HKCAAVQ	Hong Kong Council for the Accreditation of Academic and Vocational Qualifications, Hong Kong
HoP	Head of Program, SIM University
HYCU	Hanyang Cyber University, S. Korea
IA	Institutional Audit
IAAP	International Academic Advisory Panel, SIM University
ICDE	International Council for Open and Distance Education
ICT	Information and Communication Technologies
ICU	International Christian University
ID	Instructional design
IDRC	International Development Research Center
IGNOU	Indira Gandhi National Open University, India
INQAAHE	International Network for Quality Assurance Agencies in Higher Education
IQRI	Institute of Quality, Research and Innovation, Malaysia
ISO	International Organization for Standardization
ITLA	Institute for Teaching and Learning Advancement, Open University Malaysia
IVC	Internal Validation Committee, Open University of Hong Kong
JIHEE	Japan Institution for Higher Education Evaluation, Japan
JUAA	Japan University Accreditation Association
KERIS	Korea Education and Research Information Service, S. Korea

KRIVET	Korea Research Institute for Vocational Education & Training, S. Korea
LAN	*Lembaga Akreditasi Negara*
LCDMS	Learning Content Development and Management System
LMS	Learning Management System
MATE	Master of Arts in Teacher Education, Open University of Sri Lanka
MEIS	Medical Education Information System, China's Peking University School of Distance Learning
MeK	Mongolian e-Knowledge
MEST	Ministry of Education, Science and Technology, S. Korea
METEOR	Multimedia Technology Enhancement Operations, Open University Malaysia
MEXT	Ministry of Education, Culture, Sports, Science and Technology, Japan
MFOS	Mongolian Foundation for Open Society
MNCEA	Mongolian National Council for Education Accreditation
MOE	Ministry of Education
MOECS	Ministry of Education, Culture and Science, Mongolia
MOHE/MoHE	Ministry of Higher Education
MQA	Malaysian Qualifications Agency, Malaysia
MQF	Malaysian Qualifications Framework, Malaysia
MQR	Malaysian Qualifications Register, Malaysia
MUST	Mongolian University of Science and Technology
NAAC	National Assessment and Accreditation Council, India
NACTE	National Accreditation Council for Teacher Education, Pakistan
NAEAC	National Agricultural Education Accreditation Council, Pakistan
NBEAC	National Business Education Accreditation Council, Pakistan
NCEAC	National Computing Education Accreditation Council, Pakistan
NCTE	National Council for Teacher Education, India

NDCEP	Non-Diploma Continuing Education Program, China's Peking University School of Distance Learning
NIAD-UE	National Institution for Academic Degrees and University Evaluation, Japan
OASIS	Office of Academic Support and Instructional Services, University of the Philippines—Open University
ODL	Open and Distance Learning
OEAS	OpenEntryAdmissionSystem,Malaysia's Wawasan Open University
OERs	Open Educational Resources
OHEC	Office of the Higher Education Commission, Thailand
OLE	Online Learning Environment
OLI	Open Learning Institute, Hong Kong
ONESQA	Office for National Education Standards and Quality Assessment, Thailand
OPP	Outline Program Plan, Malaysia's Wawasan Open University
OPP	Outline Program Proposal, Open University of Hong Kong
OU	Open University
OUC	Open University of China
OUCS	Open University of China System
OUJ	Open University of Japan
OUM	Open University Malaysia
OUSL	Open University of Sri Lanka
OUUK	Open University, UK
OVCAA	Office of the Vice-Chancellor for Academic Affairs, University of the Philippines-Open University
PAASCU	Philippine Accrediting Association of Schools, Colleges and Universities
PACUCOA	Philippine Association of Colleges and Universities' Commission on Accreditation
PBC	Pakistan Bar Council
PCATP	Pakistan Council of Architects and Town Planners
PCP	Pakistan Pharmacy Council
PDCA	Plan-Do-Check-Act

PDD	Program Definitive Document, SIM University
PEC	Pakistan Engineering Council
PEIs	Private Education Institutes, Singapore
PI	Performance Indicators, Open University of Sri Lanka
PIs	Partner Institutions, Indira Gandhi National Open University
PMDC	Pakistan Medical & Dental Council
PNC	Pakistan Nursing Council
PriME	Inwent's Planning, Monitoring, and Evaluation System
PRVC	Program Review and Validation Committee, Open University of Hong Kong
PVMC	Pakistan Veterinary Medical Council
QA	Quality Assurance
QAAC-UGC	Quality Assurance and Accreditation Council of the UGC, Sri Lanka
QAAS	Quality Assurance and Accreditation System
QAC	Quality Assurance Committee, Open University of Hong Kong
QAC	Quality Assurance Committee, Malaysia's Wawasan Open University
QAFU	Quality Assurance Framework for Universities, Singapore
QaP	Quality as Process
QAP	Quality Assurance Policy, Malaysia's Wawasan Open University
QAU	Quality Assurance Unit, Malaysia's Wawasan Open University
QE	Quality Enhancement
QEC	Quality Enhancement Cell, Virtual University of Pakistan
QECs	Quality Enhancement Cells, Virtual University of Pakistan
QTF	Quality Task Force, Malaysia's Wawasan Open University
RTVU	Radio and TV University, China
SAIC	Self Accrediting Institution Certification, Malaysia
SAR	Self Assessment Report
SCC	Story-Centered Curriculum

SDLME	School of Distance Learning for Medical Education, China's Peking University School of Distance Learning
SED	Student's Evaluation Division, Indira Gandhi National Open University
SEU	Standards for Establishing University, Japan
SIM	Singapore Institute of Management
SIM (IGNOU)	Self Instructional Material (Indira Gandhi National Open University)
SLM	Self Learning Material, Indira Gandhi National Open University
SME	Subject Matter Expert
SOP	Standard Operating Procedure, Virtual University of Pakistan
SOPs	Standard Operating Procedures, Open University Malaysia
SOUs	State Open Universities, India
SPM-PT	Sistem Penjaminan Mutu Perguruan Tinggi
SQS	Software Quality System
STOU	Sukhothai Thammathirat Open University
STRIDE	Staff Training and Research Institute of Distance Education, Indira Gandhi National Open University
TAP	Tugas Akhir Program, Indonesia's Universitas Terbuka
TEODL	Technology-Enhanced Open Distance Learning, Wawasan Open University
TLC	Teaching and Learning Center, Singapore
TMAs	Tutor-Marked Assignments
TQF	Thailand Qualification Framework, Thailand
TQM	Total Quality Management
UAMC	University Academic Management Committee, Open University Malaysia
UC	University Council, University of the Philippines—Open University
UGC	University Grants Commission, Sri Lanka
UGC	University Grants Committee, Open University of Hong Kong
UniSIM	SIM University
UP	University of the Philippines
UPOU	University of the Philippines—Open University

UP OVPAA	University of the Philippines Office of the Vice-President for Academic Affairs
UT	Indonesia Open University or Universitas Terbuka, Indonesia
VBI	Vertical Blanking Interval
VCAA	Vice-Chancellor for Academic Affairs, the University of the Philippines
VIS	Virtual Information System
VLE	Virtual Learning Environment
VOD	Video-On-Demand
VULMS	Virtual University Learning Management System, Virtual University of Pakistan
WAP	Wireless Application Protocol
WBT	Web-Based Training
WOU	Wawasan Open University

Foreword

Discussions of quality assurance (QA) in distance education and e-learning are fairly recent. When the first open universities were established, the University of South Africa in 1946 and The Open University in the UK in 1969, there was no discussion of quality assurance as it is understood now. The term used in the 1960s and the 1970s was "standards" which Mills (in Koul and Kanwar (eds), *Towards a Culture of Quality*, COL, 2006) defined as "objective measurable outcomes." The reference point was the conventional system where high standards were defined in terms of highly qualified faculty, adequate infrastructure and facilities, regulated entry requirements, prescribed curriculum including course duration, classroom attendance, and evaluation procedures. As the *industrial production line* features of open and distance learning or ODL (in the sense of mass production and extensive mass distribution of study materials) became more and more obvious, the criteria used to measure *standards* came to be the process of course preparation and the quality of study materials; the quality of feedback and interactivity in the form of counseling, tutorials, and assignments; and the usability (in terms of pedagogic appropriateness) of ODL for the subject concerned. Further down the timeline, as ODL operations proved to be immensely user-friendly in catering to the diverse needs of the new learner (distinct from the conventional 17-/18-year-old school leaver), who came to be seen more as a customer/learner than a student, taking a cue from the corporate sector, the terms *quality* and *quality assurance* replaced the term *standards* in discussions of ODL.

The emergence and use of the Internet and the World Wide Web led to the development of e-learning and the first online course was launched in 1984. The use of web-based programs allowed a higher level of personalization and interactivity through the use of ICTs. This led to more flexible and blended approaches, adding a new dimension to QA processes.

At the turn of the present century, we saw the emergence of the Open Education Resource (OER) movement which was based on the idea that knowledge was a public good and that technology could help share, use, and reuse it. This has created yet another dimension for QA—if OER are to be harvested freely from the web, how will the quality of content be determined?

QA has been protean, and has changed its shape to deal with the emerging developments in education. The initial approaches, in the early nineties, focused more on external verification or examination of ODL practices followed by a given institution. The emphasis then shifted to the integration of both external and internal QA measures for institutions to develop "cultures of quality." With the rise in cross-border and trans-modal ODL operations, national and then regional accreditation policies and operations were put in place to assure the quality of distance education and e-learning materials, delivery, and outcomes across geographical and cultural boundaries. A more recent shift is the focus on continuous self-improvement rather than on pure accountability.

This book is a recent and valuable addition to the discussion of QA in distance education and e-learning. It identifies the major challenges and best practices from Asian higher education institutions and shares some important lessons learnt.

Let us first take up the challenges that institutions currently face. As universities make a transition from traditional ODL to e-learning, there is a huge gap in the faculty capacity to deal with the new delivery modes. Lack of training for staff in external and internal QA standards and indicators is a major stumbling block in developing "cultures of quality." Another frequently cited challenge is that many ministries and accreditation bodies use standards and indicators that have been developed for conventional universities and do not serve the purpose of ODL or e-learning well. Most QA processes cover formal education but do not take into account non-formal or informal programs. Even in countries with well-established QA processes, there is an increasing challenge for those open universities which are becoming dual-mode by beginning to offer face-to-face provision. As open universities extend their reach to other jurisdictions, adapting their own QA processes to the new jurisdictions and the corresponding local practices is a significant problem.

But Asian institutions are addressing these challenges in their unique ways. For example, the Sukothai Thamathirat Open University, Thailand has integrated external and internal QA processes to develop cultures of quality. The Open University of Hong Kong and Wawasan Open University have adopted rigorous approval and review processes for course development by creating standard operating procedures. Hanyang Cyber University, Korea, uses ICT systems for monitoring suspicious behavior from a particular IP address and for conducting online faculty orientation. The Universitas Terbuka, Indonesia, and Peking University, China, have opted for ISO certifications. Korea's AutoEver uses a variety of e-learning tools to customize and personalize the teaching-learning process. The Virtual University of Pakistan builds quality into student assessment whereas Japan's Kumamoto University, India's Indira Gandhi National Open University, the Open University of China, and the University of the Philippines Open University adopt a mixture of instructional systems design strategies and support services to develop and implement quality ODL programs.

What are some of the lessons that we can draw from the Asian experience? It is difficult to create a culture of quality through a top-down imported process. This can only take root when the staff concerned takes ownership of the processes. The more the levels of oversight set up for implementing QA, the lower the extent of faculty ownership. The responsibility for quality needs to be situated as close as possible to the operational end of a given process, as we have seen in the cases of Singapore's SIM University and the Open University Malaysia. In the absence of national QA systems, partnerships with international nonprofit organizations can help keep abreast of tried and tested QA practices as in the case of Mongolian e-Knowledge. In countries where there are QA policies in place at the national level, there is a greater likelihood of a good QA environment down the line at the institutional level as observed in the Open University of Sri Lanka. Further, the top institutional leadership has an important role to play by championing the cause of QA, providing the necessary resources and training for the staff.

It is paradoxical that even though there is a huge growth in open distance and e-learning provision, with 70 open universities in Asia alone, the issue of credibility continues to haunt the sector. As higher education institutions are called to account for their outcomes and continue to grapple with the quality of their provision, this book is a timely intervention that will take the debate on quality to the next level.

Asha Kanwar
President and CEO of The Commonwealth
of Learning (COL) in Vancouver, Canada.

REFERENCE

Mills, R. 2006. "Quality Assurance in Distance Education—Towards a Culture of Quality: A Case Study of the Open University, United Kingdom (OUUK)," in B.N. Koul, & A. Kanwar (eds), *Towards a Culture of Quality*, pp. 135–148. Vancouver: Commonwealth of Learning.

Preface[1]

While correspondence education has existed for more than 200 years, modern distance education involving the extensive use of technology is generally accepted to have started with the establishment of the Open University, UK (OUUK) in 1969. Following the success of the OUUK, a number of Open Universities were established in Asian countries, such as the Korea National OU (1972), Allama Iqbal OU (1974), Sukhothai Thammathirat OU (1978), OU Sri Lanka (1978), OU China (formerly China Central Television and Radio University, 1979), OU Japan (formerly the University of the Air, 1981), Universtas Terbuka Indonesia (1984), Indira Gandhi National OU (1985), and OU Hong Kong (formerly the Open Learning Institute, 1989). A few more were set up in the 1990s.

The period 1970 to 1990 saw the Asian OUs focus on extending access, without much regard being placed on quality. This led to concerns that the quality of education delivered through the distance mode may be inferior to its face-to-face counterparts. From the 1990s through the early 2000s, the advent of the Internet, the World Wide Web, and other communication technologies all contributed to a new global networked environment that spawned new ways of delivering education from a distance that is now referred to as e-learning.

The 1990s also saw a new trend in many Asian countries, whereby governments, realizing that they are unable to provide sufficient resources to meet the massive increase in demand for

[1] This book is funded by the IDRC's Openness and Quality in Asian Distance Education project. We are indebted to the project leaders Naveed M. Malik, Gajaraj Dhanarajan, and Maria Ng for their support and input. We extend our special thanks to Chen Li, S Baigaltugs, and Patricia Arinto for their assistance in case selection and chapter edit, to David Murphy for his undertaking of the final editorial arrangements, and to all contributors of the individual cases.

higher education, began liberalizing education policies to allow for the setting up of private colleges and universities. Inevitably, concerns were raised about the quality of education delivered by these private institutions, irrespective of the mode of delivery. Many countries began to set up regulatory and/or accreditation bodies and develop quality assurance (QA) policies for their higher education sector, including distance education (DE)[2].

Between 1990 and 2005, national level accreditation agencies/ councils were formed in Hong Kong (1990), India (1991), Indonesia (1994), Malaysia (1996), China (1999), Korea (2001), Sri Lanka (2003), Singapore (2004), and the Philippines (2005). In most cases, the agencies cover both face-to-face and DE institutions. Meanwhile, at the regional level, the Asia Pacific Quality Network (APQN)[3] was established in 2004, and immediately undertook a regional project on QA in DE. In 2008, the ASEAN Quality Assurance Network[4] was established to promote and share good practices of QA in higher education in Southeast Asia.

DE is enjoying phenomenal growth in Asia. Besides the already established OUs, a growing number of conventional institutions are offering e-learning, while new public and private DE providers as well as for-profit providers are also entering the market at a rapidly growing rate. In addition, the adoption of blended learning methods is blurring distinctions between classroom and off-campus learning, and there is also a surge in both the export and import of DE. Accompanying this expansion and increased dependency on DE to provide higher and continuing education is the growing public concern for the quality of DE. Responding both to the public demand for accountability as well as staff aspirations for quality, many established DE institutions have developed QA systems for DE, while others are at various stages of testing and development.

This book documents the existing regulatory framework covering QA in higher education in a number of Asian countries.

---

[2] Distance education (DE) in this book includes education carried out via various types of media and technologies, and e-learning. Throughout the book, DE and open and distance learning (ODL) are used interchangeably.
[3] http://www.apqn.org/
[4] http://www.mqa.gov.my/aqan/

The chapters reveal how 16 DE providers/programs have developed their QA systems and procedures to address the regulatory requirements. They present good practices in QA for DE/e-learning, analyze challenges in assuring the quality of DE products and services, offer possible solutions to meet those challenges, and present lessons for other DE providers. The 16 institutions are:

- four mega open universities (India's Indira Gandhi National Open University, Open University of China, Thailand's Sukhothai Thammathirat Open University, and Indonesia's Universitas Terbuka);
- two dedicated public open universities (Open University of Sri Lanka and University of the Philippines Open University);
- four dedicated private open universities (Malaysia's Open University Malaysia and Wawasan Open University, Open University of Hong Kong, and Singapore's SIM University);
- two online programs offered by conventional universities (China's Peking University and Japan's Kumamoto University);
- two new virtual universities (Korea's Hanyang Cyber University and Virtual University of Pakistan);
- an NGO (Mongolian e-Knowledge); and
- a for-profit corporation (Korea's AutoEverSystems).

The above selection of cases is not exhaustive, but ensures that a wide range of QA systems and perspectives of quality in DE is covered in this book.

The book is divided into five parts, depending on the unique quality focus of each case.

- Part 1 includes four cases, which highlight a balanced or systems quality assurance approach: Singapore's SIM University, Thailand's Sukhothai Thammathirat Open University, the Open University of Hong Kong, and Korea's Hanyang Cyber University.
- Part 2 includes four cases that focus on ensuring the quality of management processes: Indonesia's Universitas Terbuka,

China's Peking University, Mongolian e-Knowledge, and Korea's AutoEver.

- Part 3 includes four cases that focus on instructional design and pedagogy: Japan's Kumamoto University, Open University of China, India's Indira Gandhi National Open University, and University of the Philippines Open University.
- Part 4 includes two cases that examine QA processes covering learning support: Malaysia's Wawasan Open University and Virtual University of Pakistan.
- Part 5 includes two cases that focus on outcomes, and link QA efforts and performance measurement: Open University of Sri Lanka and Open University Malaysia.

A final chapter provides a conclusion, drawing together lessons from the cases, and looks to the future with consideration of policies required to advance quality assurance.

This book is edited by three experienced DE practitioners who have conducted a project entitled *Quality Assurance Models, Standards and Key Performance Indicators for ICT-supported Distance Education in Asia,* funded by the International Development Research Centre (IDRC)'s *Openness and Quality in Asian Distance Education* project. The contributors of the 16 case studies are academics, policy makers, and professionals with high levels of expertise in investigating and managing QA in DE.

We hope that readers will enjoy exploring the variety of innovative approaches taken to develop and improve QA systems in DE and find some useful insights for their own QA context.

Insung Jung, Tat Meng Wong,
Tian Belawati

PART I

A Systems or Balanced Approach to Quality Assurance

Singapore's SIM University[1]

Cheong Hee Kiat

INTRODUCTION

The Singapore tertiary education landscape consists of two sectors—the first comprises polytechnics, specialized institutions, and universities set up and funded by the state, and the second made up of private local and foreign institutions. The publicly-funded universities comprise the National University of Singapore, Nanyang Technological University, Singapore Management University, and the Singapore University of Technology and Design (there is a fifth institution called the Singapore Institute of Technology that may be classed together with the four in the present paper). The private sector universities are those foreign universities that have set up campuses in Singapore, or offer their programs through private education institutes (PEIs), which are run as businesses. SIM University is counted in the group of private universities.

The public and private sector institutions are subjected to different regimes of quality assurance (QA) and statutory oversight. In the former, the Ministry of Education (MOE) generally has oversight of the governance, management, education offerings, student intake, QA, and financing of the institutions. This is more so in the sub-university institutions. The public universities are autonomous (called Autonomous Universities, AUs) and enjoy more latitude in deciding their strategy and operations. Nevertheless, they are accountable to the MOE for the use of public funds, and there is thus a significant measure of oversight by, and interaction with, the MOE. These AUs are subjected to a 5-year cycle of quality audit, under the MOE's Quality Assurance Framework for Universities (QAFU). The audit is performed by an independent panel of local and international academics and

---

[1] This document was prepared with input from various individuals in UniSIM to whom the author is grateful. The information is correct as on 1 July 2011.

industry representatives appointed by the MOE. They perform whole-of-institution audits focusing on governance, management, teaching and learning, industry relations, service to society, and research. Although SIM University is a private university, it is also subjected to QAFU audit once every five years.

Within the universities, other audits are conducted to satisfy the requirements of professional bodies that accredit specific degree programs, such as that conducted by the Engineering Accreditation Board of the Institution of Engineers, Singapore, under the terms of the Washington Accord.

In addition, the MOE has appointed an International Academic Advisory Panel (IAAP) that makes recommendations to the Singapore Government on strategic matters related to the AUs. One major focus of the panel, comprising academics from leading international universities, is academic quality.

The private institutions, on the other hand, are subjected to the oversight of the Council for Private Education (CPE), with which all such institutions must be registered. The focus is less on quality than on safeguarding students' other interests such as fee protection, minimum deployment of resources for teaching, minimum qualifications of teachers, proper documentation, and so on. The academic quality is largely the concern of the university partners of the PEIs themselves.

SIM UNIVERSITY

SIM University (UniSIM) is Singapore's first privately-funded not-for-profit university. It was established in 2005, but its history of university education dates back to 1992, when the Singapore Institute of Management (SIM) collaborated with the Open University UK (OUUK) to offer OUUK degree programs. This program at SIM was subsequently approved for operation with university status.

Whilst the AUs provide undergraduate education mainly for fresh school-leavers, UniSIM offers undergraduate programs only for working adults and adult learners who need to balance their work, family responsibilities, and studies. UniSIM sees itself as a provider of education for those who have missed out on a university education. The university honors this "second chance" philosophy by admitting applicants who satisfy a minimum set of criteria, and graduating those who have met stringent graduation criteria that are compatible with the other Singapore universities. In that sense, UniSIM is not an "open" university, nor is it a distance learning institution—its delivery of education is through a mix of face-to-face sessions, comprehensive self-learning materials, and e-learning.

Accordingly, its Vision and Mission statement, and the core values, are as follows:

- Vision: *Serving society, through excellence in flexible learning for adults.*
- Mission: *To provide opportunities for professionals and adult learners to upgrade their qualifications, knowledge, and skills through a wide range of relevant programs.*
- Core Values *(SPIRIT): Spirit of learning, passion for excellence, integrity, respect and trust for individual, innovation and teamwork.*

As a private university, UniSIM is registered with the CPE. While UniSIM is not state-supported, eligible adult Singapore citizens and permanent residents enrolled at UniSIM enjoy up to 55 percent financial subsidy of their tuition fees from the Government when pursuing their first degree.

At the end of 2010, UniSIM had an enrolment of just over eleven thousand students in more than 40 degree programs, organized in four schools. The projected steady-state enrolment by 2015 is around fourteen thousand students. UniSIM has 65 full-time faculty members and nearly two hundred administrative and technical staff members. Additionally, its teaching function is supported by associate faculty from a pool of more than six hundred academics.

UniSIM is governed by a Board of Trustees comprising members from academia, industry, SIM, and MOE. The Patron of UniSIM is the President of the Republic of Singapore. As shown in Figure 1.1, UniSIM consists of the following 4 clusters:

- The President's Office, with corporate functions including the *Executive, Planning & Finance, Organizational Development, Communication, IT Services,* and the *Quality Assurance Unit.*

- The academic cluster under the provost, organized around four schools, each headed by a dean, viz., *School of Business, School of Science and Technology, School of Arts and Social Sciences*, and *School of Human Development and Social Services*. The *Office of Graduate Studies* works with the schools on graduate programs, while the *Center for Applied Research* promotes and administers the university's research activities.
- The administrative cluster is headed by the Registrar, and comprises three divisions, viz., *Admissions* (program marketing and student recruitment/admission), *Student & Alumni Relations* (overall student support & services, and relationship-building with alumni), and *Academic Services* (curriculum and examination administration, course materials management, student records, class scheduling, and appointment of associate faculty).
- The learning services cluster, led by a vice president (Learning Services), supports the learning needs of our students, and the equipping of faculty for their teaching roles. It has two arms: *Educational Technology and Production Department* (ETP)—providing printed learning materials and spearheading adoption of appropriate educational technology to create e-learning content, and *Teaching and Learning Centre* (TLC)—dedicated to raising the instructional and facilitation skill levels of our teachers.

A university-wide Teaching and Learning Committee and an e-learning Committee drive related efforts and initiatives.

Degree programs at UniSIM cover a wide range of disciplines, from business to engineering, computer technology and multimedia to languages, psychology to social work, and early childhood education to general studies. Many UniSIM programs are professionally and vocationally oriented. Most have some links with industry or other local or global tertiary institutions.

Program origination and curricular development take place predominantly within the schools, as directed and approved by the Academic Board. Each program is managed by a Head of Program (HoP) who is a full-time faculty member of a school. The dean of the school sits on the Curriculum and Planning Committee (CPC) headed by the provost. The CPC oversees the

Figure 1.1 UniSIM organization structure

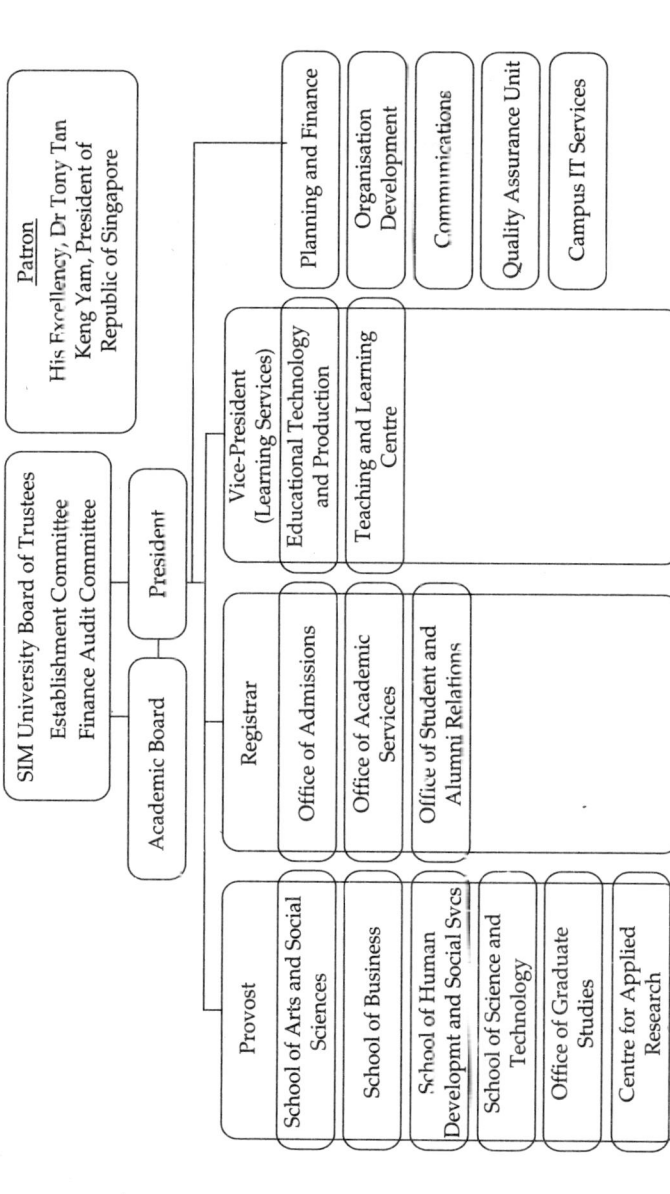

Source: SIM University website (http://www.unisim.edu.sg/about-unisim/Pages/Organisation-Structure.aspx).

development and implementation of programs and curricula. The CPC reports to the Academic Board, which in turn reports to the Board of Trustees.

Quality Management at UniSIM

UniSIM's QA system consists of internal processes, external scrutiny, and internal audit (undertaken by a Quality Assurance Unit). UniSIM is subjected to an external QAFU audit supervised by MOE. In addition, two external academic review panels, the Academic Quality Panel comprising senior academics from the local universities, and the International Academic Panel, which includes international senior academics, provide advice and feedback on current market developments, including quality measures in academia.

The university's academic standards are ensured by the following processes:

- Outcome-based programs translated into outcome-based courses, with clearly defined levels of outcomes captured in a definitive program document;
- External assessments by subject specialists for all courses;
- Outcome-based examination process incorporating monitoring, moderation, and the oversight of an external examiner (EE) for each program.
- Annual review of teaching and learning, including student feedback, AF feedback, student progression, and achievement; and
- Periodic internal and external academic audits of programs.

We need to recognize that our students have a wide spectrum of academic abilities and experience, want programs relevant to industry, are busy adults studying in a self-learning mode with multiple pathways to progress, and need a flexible and supportive learning environment.

An internal academic audit process prescribes that, each year, one school will undergo internal audit of all of its programs by an external panel, while the other schools have two of their programs audited. A quantitative measure of program quality has been introduced, an aggregate index formed by a weighted average of both course-level quality criteria and program-level quality criteria.

Academic Quality Assurance Framework

UniSIM's academic QA framework applies to the whole academic value chain, from admission through delivery of program to graduation. This is shown in Figure 1.2. The ultimate authority for all academic matters lies with the Academic Board, chaired by the president. The assessment of students is regulated through the Examinations and Awards Committee, which reports to the Academic Board. At the schools, School or Program Advisory Committees advise on directions taken or on the programs. The counsel of adjunct faculty, who are experts in their fields, is also sought.

The various components of the quality value chain are described in the following sections.

Relevant and innovative program curricula: The basic UniSIM bachelor's degree program is modular, consisting of 130 credit units (CU) of courses; for an honors degree, 40 more CUs are needed. Other programs, e.g., the Bachelor of Electronics program, are direct honors requiring 170 CUs of studies. Limited credit exemptions are given for suitable entry qualifications.

The slogan we use for our programs is "learn today, apply tomorrow," enabling our working adults to bring their learning to their workplace, sometimes applying it immediately. Market relevance is achieved through various means, including:

- market orientation of content;
- input from industry;
- content and assignments/projects with extensive industry components;
- regular curriculum reviews;
- having many AF with relevant industry experience;
- joining with industry to provide learning; and
- aligning curricula to professional accreditation requirements.

Figure 1.2 UniSIM's academic quality assurance framework

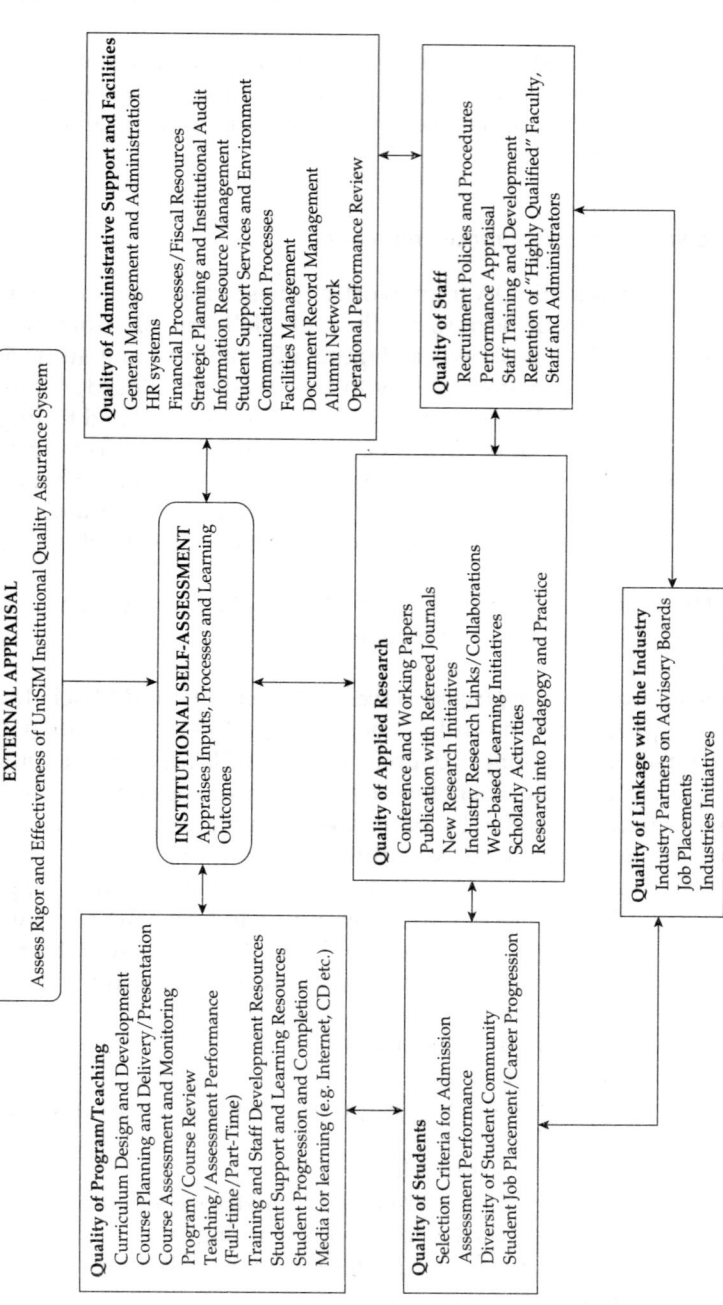

EXTERNAL APPRAISAL

Assess Rigor and Effectiveness of UniSIM Institutional Quality Assurance System

INSTITUTIONAL SELF-ASSESSMENT

Appraises Inputs, Processes and Learning Outcomes

Quality of Administrative Support and Facilities
General Management and Administration
HR systems
Financial Processes/Fiscal Resources
Strategic Planning and Institutional Audit
Information Resource Management
Student Support Services and Environment
Communication Processes
Facilities Management
Document Record Management
Alumni Network
Operational Performance Review

Quality of Staff
Recruitment Policies and Procedures
Performance Appraisal
Staff Training and Development
Retention of "Highly Qualified" Faculty, Staff and Administrators

Quality of Applied Research
Conference and Working Papers
Publication with Refereed Journals
New Research Initiatives
Industry Research Links/Collaborations
Web-based Learning Initiatives
Scholarly Activities
Research into Pedagogy and Practice

Quality of Program/Teaching
Curriculum Design and Development
Course Planning and Delivery/Presentation
Course Assessment and Monitoring
Program/Course Review
Teaching/Assessment Performance
(Full-time/Part-Time)
Training and Staff Development Resources
Student Support and Learning Resources
Student Progression and Completion
Media for learning (e.g. Internet, CD etc.)

Quality of Students
Selection Criteria for Admission
Assessment Performance
Diversity of Student Community
Student Job Placement/Career Progression

Quality of Linkage with the Industry
Industry Partners on Advisory Boards
Job Placements
Industries Initiatives

Source: Author.

Additionally, UniSIM develops programs in niche disciplines to satisfy a need for qualification upgrading in an industry/ organization. Examples are Malay Language and Literature, Counseling, Aerospace Systems, and Management and Security Studies for the Singapore Police Force. UniSIM's modular structure enables it to mix and match courses to develop new programs.

Accreditation is one measure of quality that assures stakeholders that programs meet external quality standards and graduates are well-prepared for the professions. Accreditation with professional bodies is sought at the earliest opportunity. Such programs include Electronic Engineering, Counseling, and Social Work. Others have within their curriculum opportunities for professional certification (e.g., SAP, Oracle, and Blue Coat in ICT), giving students an added advantage in the workplace.

Program Development and Review

UniSIM's system of program and course development is well-documented, setting a robust and rigorous development process that can be replicated by any new faculty member with minimal guidance. Key features of the program development process are as follows:

- New program triggered by industry need or market demand, together with strategic collaborator(s).
- Market research and consultative process involving industry practitioners.
- Curriculum development with an assembled team headed by an HoP.
- Program Definitive Document (PDD) is produced containing:
  - o market demand and sustainability analysis;
  - o outline program structure and curriculum;
  - o possibility of external collaboration or accreditation;
  - o program outcomes and mapping onto course learning outcomes;
  - o content and learning outcomes for all courses;
  - o modes of delivery and assessment; and
  - o resources needed and financial/breakeven analysis.

- External assessor reviews the proposed program.
- Approval by the CPC, and then Academic Board and Board of Trustees.

An external course assessor is appointed to validate the course after the first presentation. A program is up for review in 3–5 years, and course reviews are undertaken every 2–3 years. A review may result in phasing out of a course or program.

Program Delivery

To meet the diverse needs of learners, UniSIM's teaching and learning processes are anchored on the following broad principles:

- Stimulating a student-centric learning environment for active participation;
- Meeting the diverse needs of students;
- Engaging students to share their work experiences and relating their learning to these experiences;
- Courses designed for independent learning and reflection; and
- Various delivery modes for flexibility and to suit different course needs.

Most courses are delivered in seminar style. Comprehensive texts and study guides are supplied, augmented by face-to-face seminars and online learning interaction. Increasingly, recorded lectures/seminars, online discussions, peer-learning activities, and self-assessment quizzes make up part of the learning journey of the students. Tutor-marked assignments (TMAs) enable students to keep up with their courses and learn from tutor comments. Individual and group projects are peppered throughout a program; one constraint voiced about the assignments and projects is the required quick turnaround time for many of them. However, this mirrors real-life expectations at the workplace, and student stress levels are generally a function of the number of courses taken each semester.

As UniSIM partners with external providers to deliver some of the courses, good quality control and the alignment of objectives and standards are prime concerns in these partnerships. While our full-time faculty teaches in some courses, the bulk of the teaching is done by our Associate Faculty (AF). QA of this critical resource is vital, as described in the next section.

Associates

UniSIM has more than six hundred AF members on its active list. They are academically qualified, usually with industry experience, and comprise a healthy mix from industry and higher education institutions (HEIs). A high degree of quality control is needed to assure consistency of delivery of teaching to our students. This is through careful selection, adequate preparation and training, close monitoring, as well as discontinuation of service when performance is poor.

Recruitment policy and procedures are well established— applicants are interviewed and required to make a short teaching presentation; checks are made of original supporting documentation; successful ones are trained before appointment; and appointment is approved by the provost on behalf of the Academic Board. UniSIM deploys only teachers who have a qualification that is a level higher than the program being taught.

The training an AF member undergoes includes compulsory orientation and training in how to teach the UniSIM student, use of "MyUniSIM" (UniSIM's learning management system), online grading and detailed feedback on assignments, plagiarism, setting of assessments, and course requirements.

Attracting and retaining high caliber AF members is vital to the continued teaching standards at UniSIM. Measures, such as Teaching Excellence Awards, have been introduced to differentiate and reward exemplary AF members, while increased university interaction with them and opportunities for professional enrichment outside their teaching engagements are pursued.

E-learning

A central theme of UniSIM's program offering is "Quality Education, Anytime, Anywhere." This is still work in progress, as it involves the implementation of a full e-learning system that will deliver teaching and learning in e-mode as the main vehicle rather than as a supplementary mode. In addition to comprehensive printed texts and study guides, increasing efforts are being made to enable students to access learning through an e-mode. Present efforts include management of educational content, deployment of e-learning content, encouragement of participative learning through discussion boards, and enabling of formative assessment.

The key question is: "What is good quality in an e-course?" This is still being debated in UniSIM, but ETP has proposed a scheme to assure quality of e-courses and formally evaluate and quantify the degree of learning enhancement and learning effectiveness of these courses.

Assessment

The following widely accepted attributes of good assessment practice are adopted:

- Assessment methods (formative and summative) are valid;
- Reliability is ensured by internal moderation and external scrutiny;
- Assessment is criterion-based and addresses learning outcomes;
- Students are provided with the assessment criteria and with timely formative feedback;
- Assessment methods are designed to reduce opportunities for plagiarism and cheating; and
- Student guidance on these issues is provided.

A multi-mode assessment regime is applied to our courses. This comprises:

- Continuous assessment (CA), consisting of TMAs, quiz, online discussion (graded and ungraded), and laboratory report.
- Summative end-of-course assessment, consisting of End-of-Course Assignment (ECA), examination, project, and oral presentation (for projects, including capstone projects).

To pass a course, students must pass both continuous and end-of-course components. All CA and ECA work is submitted online and vetted by anti-plagiarism software.

UniSIM follows a 5-point Grade Point Average (GPA) system, and a minimum program GPA of 2 is required for graduation. An academic progression scheme helps students to track their progress and ability to complete a program. To avoid students lingering in a program when they are not faring well, the scheme identifies them early (academic termination for two consecutive GPAs below 2). However, they may be allowed to restart the same program or another but as new entrants.

Partnerships in Programs

One challenge is how UniSIM can provide programs over a comprehensive range of disciplines with such a small complement of full-time staff. Further, how can its programs be quickly recognized? To meet these challenges, the university encourages and supports collaborative partnerships for program/course development and delivery. Hence, UniSIM's capability and responsiveness to market supply and demand are augmented while deriving innovative teaching processes from partners. Quality and image are also enhanced through industry associations and collaborative program offerings. One example is the university's aviation programs with Cranfield University of Technology, UK and Embry-Riddle Aeronautical University, USA.

Supporting Infrastructure

Faculty & AF

Our AF members are given all the professional, administrative, and technical support they need to improve the quality of their teaching. An Associate Portal provides much of this support. UniSIM supports the continued professional development of its full-time faculty members. An annual budget is available for use by each faculty member for professional development, conference attendance, and for research purposes.

Library

An annual budget is provided for procurement of books and other resources for the campus library, but for our students and with our emphasis on e-learning, an increasing part of the budget is spent on e-materials and online databases.

Technology

UniSIM uses technology extensively to enhance learning, teaching, and the student experience. Considerable resources are devoted to IT infrastructure, bandwidth, and "MyUniSIM", which is the conduit for a number of vital teaching and learning functions, including content management, online assignment submission, assessment and academic interaction, and communications.

MyUniSIM

MyUniSIM is just one component of the computer-based support system referred to as the Student Portal. The portal provides each student with a calendar for all enrolled courses, an individualized curriculum plan, electronic course feedback, the email system, and other relevant applications.

Student Support

For our adult learners, support services are important to facilitate their studies at UniSIM. Helpdesk and student enquiry support is presently provided on a 16-hour basis. Customer relationship management (CRM) software helps staff deal with enquiries and requests, with target response times. Class and assessment scheduling and an individual curriculum plan are given to each student. A Counseling Center with professional services helps staff and students cope with various personal and work issues. Academic counseling is provided separately by HoPs.

Feedback

Various channels of feedback are utilized to gauge our provisions with a view to introducing improvements. The main channel is student feedback on teaching, e-learning, programs, services and infrastructure, undertaken every semester. Results are fed back to the relevant schools/departments for follow-up. Timely corrective action is a key focus of attention. Other channels include EE reports, input from advisory committees, graduate employment surveys, and employer surveys.

Benchmarking

While UniSIM inherited the model that served the SIM–OUUK collaboration, this has evolved as the University benchmarked itself against the local universities. UniSIM also benchmarks closely against relevant reputable overseas universities, which are either open universities, or catering mainly to adult learners on self-learning mode or are more "open access" than traditional universities. These include OUUK, Open University of Hong Kong, Athabasca University, University of Maryland University College, Fern University, and Open University of Catalonia. Visits to other institutions, and sharing with visiting experts, are means to adopting good practices. Additionally, some programs adhere to standards and requirements of professional accreditation bodies or programs in reputable universities. Being a private university, we also benchmark generally with industry/business practices of financial prudence, performance management, and the like.

In benchmarking, UniSIM needs to develop its own set of performance indicators that help it keep track of quality. This is not straightforward, and even the choice of universities to benchmark against is not obvious, given the unique nature of UniSIM.

Main Issues Faced and Solutions Adopted

UniSIM is only six years old as a full-fledged university, so its sharing of experience and lessons on the quality path is limited. The main issues pertaining to its quality management and provisions for quality education are summarized here (not in order of priority).

General Issues

Start-up and transition: One major factor that affected UniSIM's operations is that at start-up, UniSIM inherited the OUUK academic system. Establishing its own unique system of governance and operations ethos, policies, and processes was an imperative but a major undertaking. We had existing students to cater for while accepting new students into our own academic system. Hence, experienced leadership was sought to lay a good foundation.

The purpose and mission of the University had to be clear, both within the University and for external consumption. Buy-in of existing and incoming staff was critical, especially for the quality agenda. This was achieved through consensus-building and clear leadership direction.

Reputation: Several factors work against the establishment of high reputation for the University—its newness, spread of student intake quality, the "second chance" university label, adverse impressions about a university education that relies on self-learning and more-than-normal reliance on e-learning. This takes time, but expectations were high and had to be managed, especially for our staff and student stakeholders. With too many imperatives and a small staff and budget, branding and marketing were not a priority in the early years. Nevertheless, UniSIM appears to be gaining healthy acceptance by Singapore stakeholders, evidenced by recognition by the MOE of UniSIM's programs for teacher advancement, the MOE fee subsidy, and the many partners who have joined us. We need to strengthen relations with employers to get more recognition for our degrees, building more CET ties and research collaborations with companies.

Quality Assurance Issues

Our students: Understanding the needs of our adult students is pivotal to providing the right quality and platform of education. This was not immediately achieved, due to relative inexperience in dealing with a large cohort of such students. We had to evolve our policies and processes to accommodate student needs— to do this, we had to be flexible, consultative, and responsive, sometimes experimenting and willing to make changes when policies were not effective. For the latter, communication with staff was essential so as to maintain confidence in the decision-making of the leadership.

Having a "wide door, narrow exit" admission policy results in a broad spectrum of student academic abilities and performance—it was a major challenge to deal with this. Students tried to remain in the system despite not keeping up, so a stringent GPA and early termination was instituted (see earlier section on "Assessment") to prevent anyone from lingering unproductively. Much more needs to be done to help students succeed, especially those at the margins. We are only beginning to do this, by providing better support, bridging help, and counseling.

Flexible learning: Our modular system allows students to progress at their own pace. This was welcomed, but presented various challenges to the University. For example, tracking of student progress was difficult with the old legacy system UniSIM inherited. Also, each student needed to be provided with an individualized curriculum plan, but as curriculum modifications were made, ensuring updated curriculum plans became increasingly time-consuming and needed laborious checking. Predicting demand correctly for each course in a semester became increasingly difficult as students did not progress in a cohort basis, and their course choices varied according to their own circumstances. As the number of courses grew, this affected the provision of teaching resources to cater to unpredictable demands. Presently, we are developing a model to predict the demand for courses every semester.

E-learning will be a major feature of our program delivery. Various hurdles were encountered in providing quality e-learning—some faculty members were unfamiliar with, or resistant to, e-learning; students preferred face-to-face teaching; and good technical resources to ramp up e-module production were in short supply and hard to retain. When inferior quality e-learning was deployed, it caused harm to the effort and reduced student receptiveness. Another problem was the poor bandwidth that students have in their own computer systems. From UniSIM's limited experience, having a clearly enunciated e-learning strategy is critical for success, accompanied by a judicious deployment of technology, leadership to drive the effort, persuasion of staff and students, and training and assistance to enable these stakeholders to adopt, and adapt to, online learning.

It is better to roll out well-designed e-modules at a slower pace than rush in to experiment.

Staffing: There was a need to rapidly expand the number and quality of staff, particularly faculty, IT, and learning development staff for an organization that has a lean set-up. We could not recruit as fast as we wanted, even at market rates. Faculty members have to be carefully chosen—we lost a few who did not fit into the role as academic–administrators in a teaching-oriented university, and that was disruptive to our progress. New faculty members need to be well-briefed and mentored to enable them to quickly understand and adjust to our academic philosophy and operations.

As student numbers grew, employing associates in sufficient numbers, and with good qualifications and the ability to handle adult students in teaching sessions, was a major and continual undertaking. As they come from all types of background, there is a real issue of inconsistency in standards of delivery, assessment, and supervision of students—we have mandatory training before deployment, regular monitoring of teaching feedback, as well as curtailment of teaching contract if quality remains poor. A minority provides unreliable service, especially when their own full-time career interests have priority—quality is then compromised. To combat this, we have built and retained a growing team of associates who have been teaching for a number of years. Retaining good associates is a priority and various initiatives are in place to do this.

New staff have also brought in different paradigms, mindsets, and work culture, requiring adaptation all-round, sensitive staff management, and new staff policies. In particular, new faculty members who come with the mindsets of traditional HEIs must adapt quickly to UniSIM's open-access self-learning features. There is thus a need to ensure the right faculty is hired, and a thorough orientation of the new faculty to UniSIM's mode of operation and quality of delivery is given.

Resourcing: A quick-start, cost-efficient model for our growth in the early years was the use of leased facilities to cope with enrolment increases. This would have posed problems with the quantity and quality of off-site teaching facilities, except that we managed to make lease arrangements with the local polytechnics (which were well-endowed). For a long-term solution, new buildings for our campus are being built. However, private universities like UniSIM have to be mindful of being asset-light and efficient, and strike a good balance between owning and leasing facilities. This lean model of resourcing presents efficient and cost-effective operations but requires UniSIM to keep a keen eye on quality and supply of teaching resources. This is constantly being addressed as part of the QA process.

Processes and capacity to cope: In the early years, UniSIM introduced many changes at a rapid rate—these were in fundamental policies and operations, and growth in enrolment, number and variety of programs. The existing staffing and support systems were not expanding in tandem, and were heavily stressed. Communication of changes was not always complete or timely, leading to lapses in implementation, and ultimately, poor student experience. There was the occasional overstretching of our academic reach and dropping of quality due to rapid start-up of programs without the appropriate academic resources. Stronger process discipline and communication have since rectified much of the problems.

Quality culture: Instilling a quality culture in the whole university is a continual challenge, particularly with the influx of many new staff members. The heavy workload as we expanded and changed without fast-enough staff growth, and the prevailing work norms of some, meant there were some quality lapses and untidiness. As we stabilized, the focus on quality has sharpened; articulation by management, training, and internal quality audits have helped to instill a stronger quality culture.

Leadership succession: UniSIM has a lean management structure, so initially there was insufficient depth in leadership and succession plans. Sudden or inappropriate changes to leadership can have an adverse impact on quality culture and quality management; hence, strengthening of the first and second line leadership is necessary. Toward this end, a structured development program for managers and key appointment holders is being designed. For other staff members, a training and development framework, following a competency analysis exercise, is being put in place. Part of the agenda is strongly focused on quality provision.

Future Challenges

UniSIM faces several major challenges in the next few years:

Competition, demographics, and student quality: The higher education landscape in Singapore is changing rapidly. The CPE oversight of the private education sector is starting to produce a stronger sector, with major players who will further challenge UniSIM for students. The Singapore Institute of Technology started operations in 2010, and will rapidly ramp up intake of polytechnic graduates for university-level upgrading, the very student pool from which UniSIM draws. More private institutions are offering shorter degree programs and attracting students from others, such as UniSIM, that maintain a longer candidature for rigor. The outcome can mean fewer or poorer-quality applicants for UniSIM. The challenge is to attract a larger slice of the better students. Furthermore, with the declining fertility of our citizens, the number of students in the medium- and long-term for the universities is expected to decline, intensifying competition at a time when more university places are coming on stream.

Growth constraint and ability to adapt: Continual adaptation is vital for UniSIM as it increases its openness of access to aspiring students, especially for mature students. Greater openness inevitably results in the admission of a wider spectrum of academic abilities. Hence, UniSIM's capability and capacity to provide the relevant student academic support need strengthening.

Apart from the challenge of being able to recruit suitable full-time staff in the right number and quality, questions have continually been asked about the long-term sustainability of our low-cost model of teaching by associates, and also whether it is compatible with quality education. Retaining good associates and keeping them up-to-date and trained as pedagogy changes will be a challenge, especially as more and more of them are sought by the PEIs and. increasingly, by the public universities and polytechnics. The future constraint on growth of UniSIM may thus be limited by the growth in the number of associates.

Providing quality e-learning: A major challenge is the provision of high-quality e-learning. The problems UniSIM presently faces in this area, mentioned above, will remain—getting staff and student buy-in, producing quality e-materials at a quick rate, getting staff trained in sufficient numbers to function well in e-space, and having the means to evaluate the effectiveness of e-learning we provide. Securing sufficient technical and pedagogical expertise to produce or support e-modules will be a difficult problem to address.

Innovation and product differentiation: This is a perennial challenge as we seek to offer innovative and niche educational products. Our primary niche, if there be just one, is "industry oriented," "employability value-added" training. The challenge is building up our capability to read employment trends, to recognize quality-of-life enhancement niches, and to flesh out emerging generic problem-solving, communication and life skills, in order to create products which add value beyond providing a degree award.

Evolving the UniSIM brand and model: While UniSIM has progressed in its journey of becoming recognized as a high-quality educational institution, there are still misconceptions about what UniSIM caters to and what education it provides. UniSIM has some way to go in establishing its name firmly in a crowded market. It must evolve its own unique brand in the local educational landscape, increase awareness and recognition of its high-quality education provisions, strengthen its links with industry, highlight the successes of its alumni, and make a social impact through other areas such as research and service to the community.

CONCLUSION

A new university will take time to establish its reputation. In the first six years since its formation, UniSIM has firmly placed a priority on the quality of its people, processes, programs and provisions. This is a never-ending journey to keep improving and raising the standards of all that we do. Nevertheless, the key components to achieve this are in place. Ultimately, our graduates will prove the value of our programs.

Thailand's Sukhothai Thammathirat Open University

2

Pranee Sungkatavat and Theppasak Boonyarataphan

INTRODUCTION

The national quality assurance (QA) regulatory framework in Thailand operates at two complementary levels. One body sets the guidelines for the development of "Internal" QA systems and processes by the Higher Education Institutions (HEIs), while another body oversees the "External" QA that all HEIs must undertake every five years.

The Office of the Higher Education Commission (OHEC), Ministry of Education, is the main agency for establishing the internal QA policy guidelines for HEIs in Thailand (Office of the Higher Education Commission 2011). It has established a mechanism for the supervision and development of educational management for the HEIs that is both transparent and consistent with the Framework of the Second 15-Year Long Range Plan on Higher Education of Thailand (2008–2022) of the Committee for the Internal QA in Thai HEIs (Ministry of Education 2008). The OHEC has specified the components and key performance indicators for the internal QA that will ensure institutional relevance as well as responsiveness to changing contexts. The nine quality components for internal QA are laid out in Table 2.1.

The external QA process is conducted by the Office for National Education Standards and Quality Assessment (ONESQA). ONESQA is the only public organization responsible for overseeing the external QA for all HEIs in Thailand. ONESQA develops guidelines and methodology for conducting quality assessments of educational institutions by considering the goals, principles, and approach toward educational management at each

Table 2.1   Nine components and indicators for internal QA for HEIs

QA component	Indicator(s)
Corporate philosophy, mission, vision, commitments, objectives, and action plan	The procedure for implementing the institution s plan
Teaching and learning provision	- A system for curriculum development and administration - Percentages of full-time instructors with doctoral (or equivalent) degrees - Percentages of full-time instructors with academic titles - A system for teaching and learning management - A system for developing learning outcomes in accordance with characteristics of graduating students - Level of success in strengthening student ethics and morals
Student development activities	- A system for providing information and educational services to students - A system for promoting student activities
Research	- A system for developing research and creative works - A system for managing knowledge from research or creative works - Funding support for research and creative works in proportion to the number of full-time instructors and researchers in the institution
Providing community academic services	- A system for providing community academic services - Steps for delivering academic services to benefit society
Preservation of arts and culture	- A system for preserving artistic and cultural integrity
Administration and management	- Dynamic structure and status of institutional boards and administrators at all levels - Institutional development toward functioning as a knowledge-based institution - Information systems for administration and decision-making - A system for risk management - Operation in accordance with the roles and duties of the institution - Operation in accordance with the roles and duties of the institutional administrators
Finance and budgeting	- Financial and budgetary system
QA and enhancement	- A system for internal educational QA - Assessment results endorsed by the primary government agency with jurisdiction on this matter

Source: Office of the Higher Education Commission (2011).

level, as stipulated in the National Education Act, B.E.2542 (1999). Under this Act, all HEIs are subjected to an external QA once every five years, and must present the results to relevant agencies and the general public. The evaluation framework used by ONESQA includes 18 key performance indicators, grouped into 6 standards as follows (Office for National Education Standard and Quality Assessment 2001):

First Standard

Quality Standard of Graduates. There are four indicators:

1. Percentage of graduates who can secure jobs within one year, including through self-employment;
2. The quality of bachelor, master, and doctoral degree graduates in accordance with the Thai Qualifications Framework for Higher Education;
3. Number of published or disseminated articles by master's graduates; and
4. Number of published or disseminated articles by doctoral graduates.

Second Standard

Research and Creative Work. There are three indicators:

1. Percentage of published or disseminated research or innovation in proportion to the number of full-time instructors;
2. Percentage of research utilized at national and international levels in proportion to the number of full-time instructors; and
3. Number of academic works that have received quality recognition.

Third Standard

Academic Services to Society. There are two indicators:

1. Application of knowledge and experience gained from academic and professional services for improving teaching and learning processes or research; and
2. Number of academic and professional service activities/ projects responding to the needs for development and strengthening society, the community, and external organizations.

Fourth Standard

Preservation of Art and Culture. There are two indicators:

1. The total number of activities related to the promotion and support of artistic and cultural identity; and
2. Evidence of aesthetic development in the field of art and culture.

Fifth Standard

Institutional and Human Resources Development. There are three indicators:

1. Quality levels of the council of the university/institution;
2. Quality levels of the university/institution administrators; and
3. Evidence of faculty member development.

Sixth Standard

Internal QA system—Quality development, monitoring, and assessment, consisting of four indicators:

1. Internal quality assessment certified by the institution's parent organization;
2. Evidence of developments that fulfill the philosophy, mission, and goals specified at the institution's establishment;
3. Evidence of developments that affirm the points of emphasis and distinctive traits that reflect the identity of the organization; and

4. Evidence of utilized guidelines and solutions that address various social problems.

In Thailand, there is no QA system specifically designed for distance education (DE) institutions, as Sukhothai Thammathirat Open University (STOU) is the only university in the country that teaches solely via DE. Therefore, STOU is subject to the external QA evaluation instruments and standards that are designed primarily for other HEIs in Thailand. However, through increased benchmarking with distance teaching universities in other countries, efforts are currently underway to cooperatively develop a system of QA that is more suitable for STOU's open and distance learning environment.

Nevertheless, STOU has ensured that all the internal and external QA elements are integrated into an overarching framework aimed at achieving a high level of educational quality in compliance with the National Education Act, Ministry of Education's Bylaws (Ministry of Education 2010) on Criteria for Offering and Managing Degree Programs via Distance Education, B.E. 2548 (2005), the Circular on Guidelines for Offering and Managing Degree Programs via Distance Education, B.E. 2548 (2005) (Ministry of Education 2005), national quality standards, and the goals and philosophy of the University. The relationship between internal and external QA can be seen in Figure 2.1.

SUKHOTHAI THAMMATHIRAT OPEN UNIVERSITY

The Thai Government founded STOU in 1978 with a focus on lifelong learning, aiming to improve the qualifications of the

Figure 2.1 Relationship between internal and external QA measures at STOU

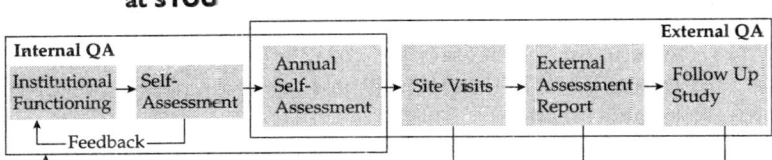

Source: Office of the Higher Education Commission (2010, p. 17).

general public through academic and professional training and increased post-secondary educational opportunities.

Over the past three decades, STOU has addressed individual and societal needs by implementing a DE system employing print as core medium, supplemented by broadcast, face-to-face tutorial sessions and e-learning, to allow STOU students to study on their own, with or without assistance from instructors or having to attend regular, conventional classes. As of 2011, STOU offers 44 bachelor's, 32 master's, and 5 doctorate degree programs through 12 schools of study covering a wide range of quantitative and qualitative academic fields. Approximately one hundred and sixty thousand learners study these programs each year.

In 1995, STOU was, out of 79 institutions worldwide, presented the Award of Excellence by the International Council for Open and Distance Education and the Commonwealth of Learning for institutions demonstrating excellence in the management of DE. The award recognized STOU's expansion of educational opportunities, development of teaching and learning techniques for DE, and support for national and regional DE. In 1996, STOU received the Asian Management Award for outstanding administration development over other educational and business organizations in Hong Kong, Indonesia, Malaysia, the Philippines, Singapore, and Thailand. This award is granted to public and private nonprofit entities that aim to improve the quality of life of the general public, and is based on the opinions of people they have served as well as an assessment by a committee of management experts.

QUALITY ASSURANCE AT STOU

The QA process in STOU addresses the following:

- Developing curricula that comply with the Ministry of Education's Thailand Qualification Framework (TQF) of higher education;
- Implementing the concept of systematic courses in the form of block courses of six credits, each combining theory with practical experience such as assigned activities, work-based

practical experience, pre-study and post-study evaluation and final examinations;

- Appointing a Course Production and Administration Team (CPAT) for each course to plan, prepare and produce course materials, and teach and evaluate the classes;
- Providing an information system allowing students to study on their own through the integration of printed core materials and supplementary media accessible to all students, such as video, audio, and electronic resources; and
- Providing supplementary media through interactive distance learning activities, radio and television broadcasts, computer-assisted instruction, e-learning, tutorial sessions, e-seminars, teleconference, and real and virtual practical experience programs.

For all of the above conditions, two dimensions of QA are implemented at STOU: internal QA and external QA. These processes are carried out according to the Thai Ministry of Education guidelines, as detailed below.

Internal QA

Offices taking responsibility for the internal QA at STOU operate at two levels: university and individual school levels. At the university level, qualified experts from within and outside the University have the duty to set QA policies and directions, and to oversee the institution's QA system. In each of the 12 Schools, a committee is set up to implement the procedures and regulations of the QA system laid out by the University. There is also the QA Coordination Center, which coordinates quality issues in all departments of the University.

All QA processes and procedures at STOU are laid out in a "QA Calendar" in each academic year. The calendar is sent out to each unit in the University ahead of time to be used in their annual operational planning. All processes are divided into four main steps, as follows.

1) Plan
   - Hold a meeting to clarify the QA framework.
   - Introduce QA concepts to the staff involved.

- Create a training program for educational quality assessors.
- Collect educational QA information and disseminate the information via the University's website and the QA Management Information System.
- The Schools and supporting agencies verify the information.

2) Do
- The Schools and supporting agencies conduct QA in accordance with the standards and quality indicators that are set up for each academic year.
- Each School and administrative unit creates a Self Assessment Report (SAR) for evaluation.

3) Check
- The Schools and supporting agencies send the SARs and lists of people with specialized knowledge of QA to the University's QA Coordination Center.
- A list of QA assessors is submitted to the University QA Evaluation Committee.
- The QA Coordination Center issues invitations to QA assessors and nominates a QA evaluation committee for each unit.
- The committee evaluates the internal quality of every School and administrative unit.
- The QA Coordination Center collects all information about QA and evaluation results from the Schools and supporting agencies to undertake a university-level SAR.
- Assessors are nominated and invited by the university QA Evaluation Committee.
- The university's educational quality is evaluated by the university QA Evaluation Committee.
- Submit the annual university QA report for the internal QA to the Office of the Higher Education Commission (OHEC) and submit the annual university QA report for the external QA to the Office for National Education Standards and Quality Assessment.

4) Act
- Hold a meeting or seminar for knowledge-sharing within the internal Educational QA Committee.
- Each unit director studies the evaluation results and suggestions from the internal Educational QA Committee

and from the University to develop a QA enhancement plan, plan operational improvement, adjust the strategic plan or annual action plan, or make quality development projects for the next academic year.

Overall, this QA system significantly benefits STOU by attesting to the quality of DE, including curricula, subject offerings, services for students, and the graduates themselves. This was confirmed by the results of a 2010 study by the National Statistical Office of Thailand, in which 95 percent of employers and other supervisors were satisfied with the quality of STOU graduates.

Integrating Internal and External QA

STOU places a high priority on QA in its education and management by applying policies that include:

- developing an education QA system that is consistent with existing internal working procedures in order to maintain the University's educational standards according to the system of standards, rules, and guidelines for quality control, quality evaluation, and QA laid-down by the Educational QA Committee;
- supporting 12 Schools, offices, institutes, centers and other divisions within STOU to improve the quality of their services by establishing related committees or working groups on issues of quality;
- promoting and supporting the participation of public and private external agencies in the University's quality assessment process in order to consistently develop and make adjustments to educational quality at STOU;
- supporting the participation of faculty, staff and students in the QA process, and promoting an awareness of all parties regarding their responsibilities in fostering the quality; and
- encouraging public relations and dissemination of information on the University's educational QA activities.

To implement these QA policies, STOU has:

- appointed the Educational QA Committee. This Committee is made up of qualified experts from both inside and outside the University, and is responsible for setting policy, making decisions, and overseeing QA in the University so that it complies with the goals and standards of STOU and the country;

- appointed QA Working Groups for Schools, offices, institutes, centers, and supporting agencies. Each QA Working Group is responsible for planning and executing QA within their respective divisions in accordance with the aims and standards of the University and the country;
- established the QA Coordination Center. The QA Coordination Center is a university-level body responsible for planning and coordinating with internal and external agencies to enable each agency to conduct their work according to the goals or standards of the University and the country;
- adopted the "PDCA" (Plan–Do–Check–Act) process as the principal mechanism for addressing each QA indicator;
- integrated internal and external QA indicators;
- established and maintained a QA database. A common data set is developed and maintained to act as a central source of information for QA. The QA Management Information System is a database to be used in QA activities, such as planning, decision-making, and preparing online QA assessment;
- developed the STOU QA Manual. The Manual integrates the national, internal, and external QA indicators into the STOU's internal system so that every section of the University is able to carry out QA under the same standards. The Manual also has a calendar showing the QA activities that every unit of the University must conduct;
- reported self-assessment results. All units create self-assessment report cards following the same format to facilitate easy comparison of assessment results;
- built the capacities of QA assessors. STOU seeks to develop the capability of QA assessors to help them check and promote QA at STOU. A QA orientation session is held each year before a formal QA evaluation begins to help assessors operate according to the same QA standards;
- evaluated the QA process. Meta-evaluation is conducted by the QA Evaluation Committee to check the QA process of each section and unit;

- used QA evaluation results for improvement. All STOU sections are responsible for creating plans for raising the level of QA in their own unit by addressing weaknesses and recommendations raised by the QA Evaluation Committee from the previous year;
- shared best practices. A process of best practice sharing is implemented through comparative studies of QA at open universities in other countries, such as the University of South Africa, Universitas Terbuka (Indonesia), Indira Gandhi National Open University (India), and the Korea National Open University, as well as other universities in Thailand. STOU is also able to share results that have proven successful in internal university units through a regular knowledge management process in the University, so that these findings can be used for continual development of STOU's QA system;
- operated a QA warning system. The University communicates important QA indicators via a QA warning system in a regular report with related units of the University. Each unit is required to operate under this system in order to make a 6- and 9-month QA report in accordance with its annual work plan; and
- awarded high-quality results. Every year, there is an award given to units that demonstrate high-quality work, in order to encourage all units to work for positive QA results.

During these QA processes, the University Council plays an integral role, especially in QA policy and strategy formulation, supervision of QA activities and follow-ups of QA evaluations.

QA Evaluation Results

In the external QA ratings by ONESQA, STOU has been rated excellent in its teaching and learning system. For example, in 2009, the overall evaluation score was 2.72 out of 3.0, which indicates recognition of STOU's excellent and exemplary educational provision. In particular, the quality of STOU's educational programs reached 100 percent of the Ministry's DE standards. The University's committee for program management and the committee for course-block management, both of which focus on working as a team to develop all programs, course materials, and instructional media have contributed to the development of a strong QA mechanism for program and course management. STOU also achieved a high-quality rating for knowledge transmission to students (100 percent), and a high-quality rating for assessment and evaluation of the DE system (100 percent). The results of this external QA evaluation confirmed that STOU has successfully addressed its core mission of making education accessible to all.

STOU's contribution to the underprivileged has been highly evaluated. For example, STOU operates a system to support prison inmates to register, study, access educational services, and take examinations, all from their place of incarceration. STOU also implements a system to help disabled individuals study using learning materials suitable for blind or vision-impaired students. STOU serves the elderly by providing training courses, special seminars, and research that focuses on the needs and concerns of the elderly community.

The quality of graduates is positively recognized by 88.4 percent of employers, entrepreneurs, and users of graduates. To date, 41 graduates have received national awards for high academic achievements in their theses. Moreover, 87 percent of students and graduates express their satisfaction with learning quality and support, and state that they are able to apply their knowledge to work at a higher level.

STOU is also recognized as a DE institution that has a concrete supporting system for creating and disseminating research projects. Results of research projects are published in STOU's e-journals and stored in an e-research management system.

ISSUES AND CONCERNS

Overall, STOU's QA system is effective in helping the institution maintain its standing as a provider of quality higher education in Thailand. However, we have identified three main challenges in both internal and external QA systems.

First, the QA standards and indicators used by the OHEC (STOU's internal QA unit) and the ONESQA (the external QA agency) are not always consistent. This is a significant obstacle, as the discrepancies force STOU to adjust its QA plan to fit with the criteria laid out by the ONESQA indicators. Being required to adhere to indicators designed for conventional universities can be impractical and time-consuming for an open university, as every inconsistency must be clarified. This extra process of explanation is the continuing responsibility for STOU.

Secondly, further to the issue discussed above, some QA indicators specified by the ONESQA for conventional universities are inapplicable in the context of DE institutions. For example, one indicator for assessing the quality of a bachelor's program is to determine whether its graduates start a job or are self-employed within one year after the graduation. But in case of STOU, where over 95 percent of students are already employed, this indicator does not show the quality of its program. Instead, graduates' application of their knowledge to their work or daily life may be a more suitable indicator for STOU. The indicator on the library, instructional media, and the learning environment is also difficult for STOU to apply. The QA standard that states that the university should have at least one computer for every eight full-time students is hardly suitable for an open university with no conventional classrooms. Another indicator states that each university shall assess learner satisfaction in teaching and learning quality and instructional support elements for every subject in each semester. This is also not suitable for a distance learning institution like STOU, where only selected bachelor's level courses include extra face-to-face instruction, and most students are self-learners.

Finally, STOU's QA system requires extensive coordination between every unit in the University, since it is designed to provide shared services for a harmonized management of all tasks. The 12 Schools focus mainly on academic content, while the supporting agencies work to support the Schools' operation. Therefore, the supporting agencies such as the Office of Academic Affairs, the Office of Registration, Records and Evaluation, the Office of Educational Services, and the Institute for Research and Development are also required to support the QA operation of the Schools. In consequence, the University QA Coordination Center is obliged to reconcile the activities of the Schools with those of the supporting agencies in order to conduct the overall University QA while at the same time complying with national objectives and standards, which has not been an easy task.

LESSONS LEARNED

Despite the challenges indicated above, STOU has been successful in developing and implementing an effective QA system. STOU's QA case provides several lessons for other DE providers in Asia.

- A QA system should integrate both internal and external QA requirements, and reflect the unique features of distance teaching and learning.
- It needs to include QA measures for developing high-quality DE programs and materials that are adaptable to formal, non-formal and informal education, covering a wide range of target groups. In providing these three types of education through DE, STOU fulfills a different role than conventional universities with clear classification of students by year, so QA measures need to reflect this difference.
- Standardized QA measures should be developed and applied to DE programs, especially for underprivileged target groups such as detainees, the disabled, and the elderly. STOU's guiding principles focus on providing education for every target group, in accordance with Thailand's national policy to expand educational opportunity, especially to underprivileged groups. QA measures focused on these learners will further the pursuit of such policy goals.
- A centralized QA unit that oversees an institution's overall QA activities should be in place in order to maximize institutional QA efforts effectively and efficiently. This unit can help facilitate unified, information-based institutional planning that can be easily followed and overseen.
- A QA manual that encompasses all QA standards and indicators in a consistent manner contributes to standardizing the QA activities of all members.

CONCLUSION

STOU strives to enhance its QA system within a conceptual framework that meets international standards. Specifically, STOU aims to apply the Malcolm Baldrige National Quality Award model (The National Institute of Standards and Technology,

Figure 2.2 Integration of a total quality management model in STOU's QA system

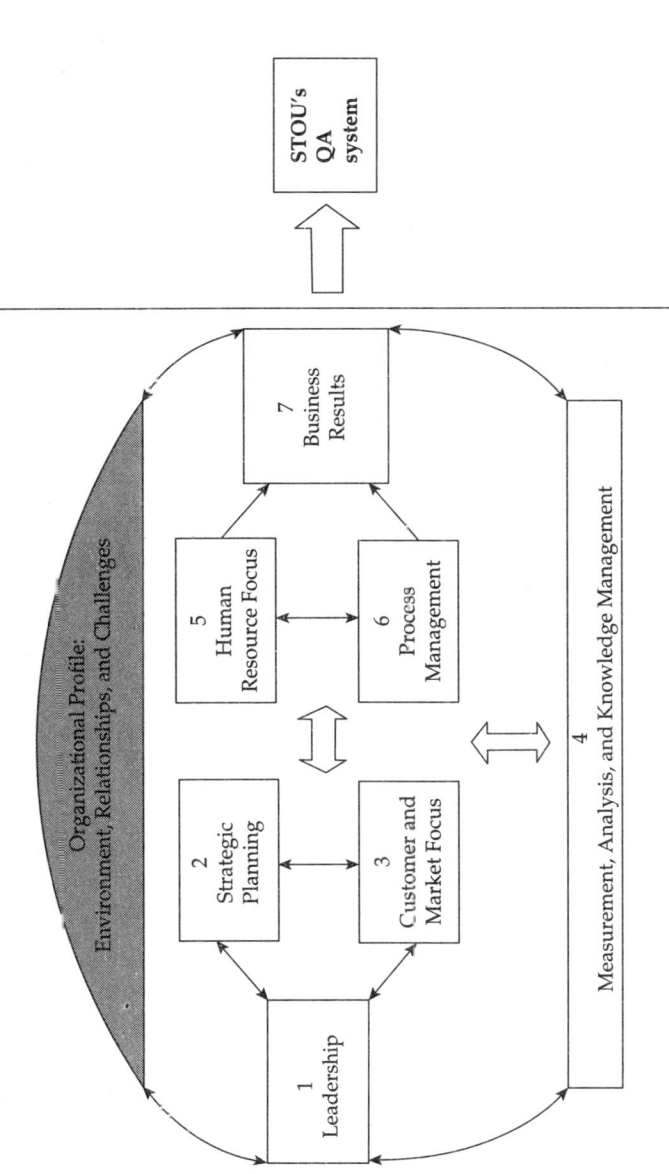

Baldrige Criteria for Performance Excellence Framework: A Systems Perspective

Organizational Profile:
Environment, Relationships, and Challenges

1 Leadership

2 Strategic Planning

3 Customer and Market Focus

5 Human Resource Focus

6 Process Management

7 Business Results

4 Measurement, Analysis, and Knowledge Management

STOU's QA system

Source: Adapted from Blazey, Mark L. (2011). *Insights to Performance Excellence 2011-2012.* Milwaukee: American Society for Quality, Quality Press.

2011) from the United States to its DE in order to build a total quality management (TQM) system. As shown in Figure 2.2, the model includes seven standard criteria: leadership; strategic planning; customer and market focus; measurement, analysis, and knowledge management; human resource focus; process management; and business result. Rules arising from these criteria are continually adjusted to the STOU's QA system.

As mentioned above, many indicators established as national standards for HEIs are not always applicable to the context of distance learning. STOU plans to develop specific standards that are better suited to DE environments, and collaborate with open universities in other countries to develop specific performance indicators for DE that can be used as benchmarks in the future.

STOU will continuously strive to promote cooperative QA activities with universities in Thailand and abroad, as well as exchange knowledge and experience with other HEIs under the QA framework of the ASEAN University Network. Moreover, STOU aims to benchmark with open universities concerning best practices in the region and around the world, particularly in such areas as supporting distance learning, producing quality graduates, using media effectively for DE, developing evaluation and assessment suitable for DE, providing educational services appropriate for DE, and elaborating a QA system.

REFERENCES

Blazey, Mark L. 2011. *Insights to Performance Excellence 2011–2012.* Milwaukee: American Society for Quality, Quality Press.

Ministry of Education. 2005. *The Declaration of the Office of the Higher Education Commission on Rules and Regulations for Establishing and Administering Distance Learning in Higher Education B.E. 2548 (2005).* Report by the Ministry of Education, Thailand.

———. 2008. *The Second 15-Year Long Range Plan on Higher Education of Thailand (2008–2022).*

———. 2010. *The Ministerial Regulation Regarding the Systems, Regulations, and Methods for Internal QA B.E. 2553 (2010).* Report by the Ministry of Education, Thailand.

Office for National Education Standard and Quality Assessment. 2001. *A Manual for External QA for Higher Education Institution B.E. 2554–2558 (2011–2015).* ONESQA, Thailand.

Office of the National Education Commission. 2003. *National Education Act E.E. 2542 (1999) and Amendments (Second National Education Act B.E. 2545 (2002))*. Bangkok: Pimdeekarnpim Co, Ltd.

Office of the Higher Education Commission. 2010. *A Manual of Internal Quality Assurance*. Bangkok: Phab Pim Ltd.

———. 2011. *A Manual for QA in Higher Education Institutions B.E. 2553 (2010)*. Report by the Higher Education Commission, Thailand.

The National Institute of Standards and Technology. 2011. *A Manual of MBNQA*. Baldrige Performance Excellence Program, USA.

Open University of Hong Kong 3

Robert Edward Butcher

INTRODUCTION

Of the ten universities in Hong Kong, the government funds eight, while two are self-financing. External oversight of academic quality in publicly funded institutions of higher education is the responsibility of the funding body, the University Grants Committee (UGC), through its semi-autonomous Quality Assurance Committee (QAC), which conducts periodic institutional quality audits. Self-financing providers are subject to institutional accreditation and program validation by the Hong Kong Council for Academic Accreditation (renamed in 2007 as the Hong Kong Council for the Accreditation of Academic and Vocational Qualifications, HKCAAVQ). In the case of the Open University of Hong Kong (OUHK), which was established by the government in 1989 but became self-financing after four years of operation, the government mandated that after attaining self-accrediting status and university title in 1997, it should continue to submit itself to an Institutional Review every five years to ensure that it was continuing to operate effectively and in accordance with its Ordinance. To date, the HKCAAVQ was contracted to carry out the review exercises, in 2002 and 2008 (OUHK 2008).

The following extract from the University's QA handbook encapsulates the OUHK's philosophy on quality:

> The aim within the University has always been to develop a "quality culture"—i.e. a collective commitment by all staff to professional excellence. This is no more than a reassertion of traditional academic values. Briefly, we believe that responsibility for quality lies with each individual and group within the University. In order to ensure the commitment of all staff to the processes, staff are given the opportunities to provide their own input to the development,

monitoring and evolution of the system. It is our belief that we should constantly aim to improve ourselves. (OUHK 2010)

OPEN UNIVERSITY OF HONG KONG

In 2009-10 the OUHK celebrated the twentieth anniversary of its establishment. It was founded as the Open Learning Institute of Hong Kong (OLI) by the Hong Kong Government to "provide opportunities for higher education by means of open learning and thereby enhance learning and knowledge and enhance economic and social development in Hong Kong" (OLI Ordinance 1989). At the time of the OLI's launch, fully funded full-time places in local conventional tertiary institutions in Hong Kong were available for only six percent of the eligible age cohort. Decades of under-provision had created a vast backlog of potential adult learners seeking access to flexible, cost-effective higher education (Hope and Butcher 2005).

When it was launched, the OUHK's distinctive features raised significant quality concerns. It operated an open entry policy that did not require applicants to have matriculated from secondary education in order to be eligible to study, creating concerns about academic standards. Students registered on individual courses, not programs of study, creating concerns about an unstructured scattergun approach to program construction. Students could study as many credits as they wished in order to meet their learning goals, accumulating credits toward defined awards over an unspecified time, raising concerns about the counting of obsolete knowledge toward an award. No penalty was imposed for taking time out from study, and it was very difficult to convince the public that high "dropout" did not necessarily represent a failure of the system. Students were not taught by full-time academic staff and, in most cases, the full-time Course Coordinators were not responsible for writing the courses, raising issues of academic ownership and standards. All students were allocated a part-time tutor. In addition to grading assignments and providing detailed individual feedback on performance, the tutor conducted tutorials in the evening and at weekends. Attendance at tutorials was not compulsory, raising concerns about plagiarism and student substitution, which were only partly alleviated by the requirement for all courses to include a compulsory examination component. Learning was achieved by interaction with the comprehensive learner-centered course materials, raising concerns about the viability of independent, self-directed learning in a culture more accustomed to a classroom-based, teacher-led pedagogy.

For these reasons, many people thought that open access distance education (DE) would never catch on in Hong Kong, but by the end of the first application period in 1989 more than sixty-three thousand applications had been received. Eventually, nearly four and a half thousand students enrolled on eight courses and the only dedicated DE provider authorized to award degrees in Hong Kong was launched (Kiloh 1999).

The institution grew rapidly, and the number of students studying for a qualification by DE reached a peak in October 2001. At this time there were about twenty-seven thousand registered active students. Since then, the number of DE students has decreased dramatically to just over thirteen thousand students studying one hundred and forty four programs in the Schools of Science and Technology, Business and Administration, Education and Languages, and Arts and Social Sciences. In October 2010 the OUHK presented four hundred and seventeen DE courses.

QUALITY ASSURANCE AT THE OUHK

The OUHK's quality agenda must always be mindful of the HKCAAVQ's QA guidelines which define the standards that the University must maintain in respect of Institutional Management and Strategic Planning; Academic Development and Program Development and Management; Staffing and Staff Development; Resources and Support Services; and Quality Assurance Systems and Mechanisms in order to satisfy the Review Panels.

At start-up, in addition to course materials, the OUHK "bought in" expertise and administrative systems (including QA systems) from the Open University, UK (OUUK) and other reputable distance learning universities around the Common-wealth (Butcher and Hope 2006). By importing its first QA systems wholesale from the OUUK, the OUHK hoped to "quickly demonstrate externally that quality systems (were) in place" (Robertshaw 1997: 67). Robertshaw points to the weakness of such a model that may result in an alienation of staff and a lack of ownership of procedures that are essentially top-down and bureaucratic, and are seen to imply a lack of trust in the academics at course-level. Following the 1995 review, where the HKCAAVQ Panel criticized the OUHK for having overly complex and elaborate QA protocols, the OUHK sought to simplify its systems, to increase participation and ownership, and to raise awareness of quality issues. By 2005, that quest had culminated in a significantly simplified QA system described by OUHK staff as effective, devolved (to School and program-team level) and learner-centered (Butcher and Hope 2006).

This case study will focus on the QA mechanisms operated by the OUHK in areas such as course and program development and approval, and assessment. It will reflect on how these mechanisms have evolved and developed over time to reflect increasing institutional maturity and changing needs.

Program Development and Approval

In the OUHK, a program is a distinct, named academic qualification. Proposals for new programs pass through a two-stage review and approval process. The first stage, outline approval, allows the academic unit (School) and the Senate to identify broad issues relevant to the offering of the program, such as marketability and its relation to existing programs. The second stage, detailed approval, allows for the thorough review of the program's academic validity, standards, and quality.

The School produces an *Outline Program Proposal* (OPP) document for approval by the Senate which contains statements of the program aims and objectives, the need for the program, language of instruction, the placement of the program on the Qualifications Framework (a Government initiated 7-level hierarchy designed to clearly define the standards of different qualifications, ensure their quality and indicate the articulation ladders between different levels), the proposed academic structure including required and elective courses, any requirements for developing new courses (because of the cost implications), projected enrolments, and a proposed development schedule.

A *Detailed Program Proposal* (DPP) is considered at the second stage and is a much more comprehensive statement of the program's aims, structure, and content, including syllabuses for each course and modes of assessment to be adopted. The DPP also includes a detailed business plan for the program. In preparing the DPP, the School is expected to seek the advice of the Advisory Peer Group (APG). This group, established for each program, includes between four and six representatives of local business, professional, and/or academic communities. It offers advice during the development of the program, and continues to meet annually to review progress once the program has been launched. The DPP is first reviewed by an Internal Validation Committee (IVC), a group comprising mainly the academic unit heads. It is then forwarded to an External Program Assessor (EPA), who is usually an academic of distinguished reputation in the field of study covered by the program proposal, most often a Professor at another local university. The EPA prepares a written report covering such issues as the academic standard of the proposed program, the suitability of the curriculum and its comparability with other, similar, local, and overseas programs. This report is then considered at a meeting of the Program Review and Validation Committee (PRVC). In addition to internal members, the PRVC has four permanent external members who are familiar with the local education scene. The EPA will also be in attendance at the meeting when the DPP is considered. Finally, the DPP must receive the approval of the University Senate, at which time the School must show how they have responded to any issues raised during the approval process. Such a system helps the University to ensure that the School has given due regard to issues raised by the external and internal reviewers and, if action has not been taken, to put forward arguments as to why changes are not considered appropriate. We believe that this system ensures that we maintain a strong internal institutional component to QA. The external members of the PVRC ensure that quality is benchmarked with practice in other universities.

Program Review, Revision, and Revalidation

All programs are subject to ongoing and regular review, and updating by the respective Program Teams. They are also subject to a formal internal revalidation process to ensure that they continue to uphold the standards of quality expected by the University. The normal period for revalidation for an honors degree is six years.

The process used for program revalidation is very similar to the process followed for the development of new programs, and is based on HKCAAVQ requirements. The program revalidation document focuses on the achievement of the aims and learning outcomes of the program and various performance indicators such as student and graduate outcomes, student and employer surveys, reports from External Examiners, and so on. The Program Team will also put forward plans for the continued development of the program. Revalidation submissions are reviewed by the APG, the IVC and the EPA. The PRVC's scrutiny of the submission will focus on whether the program has established the predicted academic validity, standards, and quality. Revalidation of a program is ultimately granted by the Senate.

Course Development and Approval

The development and production of course materials is at the very heart of the OUHK's quality enterprise. The QA processes are very stringent and, over the years, this has probably been the aspect that has created most tension between the University academics and managers.

The University's courses are largely print-based, supplemented as appropriate with audio and video components and online resources. In addition to the course materials, all courses have an online element to allow interaction between tutors and students through the Online Learning Environment (OLE). An increasing number of courses are now fully online.

A course in the OUHK is an individual subject of study and a program is made up of multiple courses. The Course Development Team (Course Team) is the body responsible for developing a new course and for ensuring that the materials meet the University's standards. Members of the Course Team include academic members with relevant expertise from the academic unit, a Course Designer (who provides instructional design input and advises on the development of educationally sound self-instructional material, appropriate assessment strategies and the use of media and technology), and a Course Developer (who may be internal or external) who is the content writer.

The University used to appoint an External Course Assessor (ECA) for each course in development. However, in a bid to simplify its processes, an ECA will not be appointed if the University has at least two independent academic staff, not involved in the development, who possess the academic expertise necessary to make a sound judgment about the standard of the materials. Where an ECA is appointed, this person acts as an independent reviewer of the academic content of the course. He or she advises the Course Team on such matters as the level of difficulty of the materials, the suitability of the assessment items, and the comparability of the course syllabus with those of similar courses elsewhere. A favorable ECA report on the completed course is necessary before a course can be approved for presentation.

Before a course is presented, it must receive approval from the Senate (on the advice of the Chair of the IVC). If the development of the whole course is not complete at the time the Senate reviews the course, Senate approval is given for one presentation only.

If the University proposes to use a course produced by another institution, such as the OUUK, we review whether the course materials are appropriate for Hong Kong students and, where necessary, we take action to supplement the materials with localized content. In such cases, the question of whether to appoint an ECA is considered on a case by case basis and depends on the issues detailed earlier, along with the availability of an external review report from the institution supplying the course.

Course Review and Revision

After the completion of each academic semester, the University reviews the presentation of courses. This review process is intended to encourage the critical, continuous review of courses with a view to their further improvement. It also ensures that feedback from students, tutors, and external examiners is taken into account.

At the heart of the procedures for the review of course presentation is the Course Report. This document is prepared by the Course Coordinator, who has overall responsibility for supervising the course. A Course Report includes discussion of any problems that have occurred during the presentation under review, notes on student performance and on feedback received from participants, and any specific recommendations for changes to the course.

The Course Reports for all courses presented in a program are reviewed by the Program Team. The Program Team is charged with making a course-by-course comparative review of student performance and of issues arising during presentation, and making recommendations to the academic unit that address both course-specific matters and concerns of more general interest via the Program Team Report. Academic units produce a summary report for each presentation that highlights major issues arising from the presentation of the unit's courses. These reports are considered at the University level by the Academic Standards Review Group, which focuses on areas of general concern and recommends appropriate actions to address them to the Senate.

Course Coordinators are expected to continually update and revise the material in their courses. Revisions to courses are classed as "minor" or "major." A minor revision is defined as one that results in changes to no more than 30 percent of the overall course content and no more than 50 percent in the content of any one study unit (courses are made up of individual units). A minor revision may be approved by the School Dean. In the case of major revisions, a new Course Team will be appointed and the provisions outlined for new course development are followed.

Student Support

While we share with other distance learning institutions an emphasis on the course materials as the main vehicle for delivering course content, our experience has demonstrated that students can encounter a range of difficulties in studying at a distance. We have developed learner support mechanisms to give assistance to our students as they need it.

The tutor is the key element of the OUHK's learning support system, providing the human interface between the University and its learners. All tutors are part-time employees of the University. The University runs mandatory face-to-face orientation and training sessions for all new tutors, which they are paid to attend. They are also provided with a self-learning course with a set of comprehensive course materials that provides general information about DE and the OU environment and generic tutoring skills. Course-specific training is also provided to all tutors (new and continuing) and attendance at these sessions is also mandatory and paid.

Each tutor is under the direct supervision of a full-time Course Coordinator. As part of the training and monitoring process, Course Coordinators will pay periodic visits to tutorials. Finally, in order to ensure that tutors do not become overloaded with work, the University restricts tutors to a maximum of two tutor groups in any teaching period.

Assessment

The integrity of assessment in a distance learning environment is of particular importance. Assessment at the OUHK is on a course-by-course basis. The University utilizes two forms of assessment—continuous assessment and an examination at the end of each course. To pass the course, students must pass in both elements of assessment.

Because the tutors provide continuous assessment, the University has set up an elaborate system for monitoring the marking of assignments. The main responsibility for this rests with the Course Coordinator, who reviews a number of scripts for each tutor, to check the quality of the marking and teaching comments, and the consistency in applying the marking guidelines. Such monitoring has the added benefit of informing coordinators whether the assignments are at the right level of difficulty and whether learning objectives are being achieved. For new tutors, or tutors who have had problems with their marking in the past, more assignments are monitored. Coordinators also monitor the "turnaround rate" for assignments, that is, how quickly they are marked by tutors and returned to students.

Tutors also mark examination scripts, and the process is subject to careful scrutiny by Course Coordinators. The University employs a complex system of marks standardization for those courses taught by multiple tutors to ensure that all students are treated fairly.

Originally, the University appointed a separate External Examiner (EE) for each course. However, as part of the QA streamlining initiative, EEs may now be appointed for a cluster of up to six courses. EEs are sent sample-marked assignments for information at the end of the course presentation. They are invited to comment on the examination question paper and review a sample of marked examination scripts. EEs must also report to the University President, providing their judgment on the performance of the students in the course and indicating whether it is comparable to students in other institutions of higher education. Areas of students' strengths and weaknesses are identified by EEs, and the Course Coordinator takes these into account when the course materials are being revised.

Each course has an Award Committee that is responsible for the award of student results. The Award Committee is chaired by a senior academic staff member of the relevant academic unit. The other members are the Course Coordinator and any other academic staff having responsibility for assessment. Since 2004, the responsibility for the approval of student course results has fallen to the relevant academic unit and no longer needs approval by the University Senate.

IMPACT OF NEW DEVELOPMENTS

Introduction of Face-to-face Provision

In the past few years, significant changes in the OUHK's operating context have intensified competition for students and necessitated change to ensure the institution's survival. In October 2001, the University took the bold step of offering the first of a series of degree programs by full-time face-to-face study, and became a player in the traditional school leaver market. From April 2002, the number of students studying by DE started to show a steady decline, until by October 2010, there were just over thirteen thousand active students. Concomitantly, the University felt that it was mature enough to take on traditional teaching modes as well as distance learning, to satisfy the largely unmet demand of school leavers to pursue tertiary education.

Having first established and then streamlined our QA protocols to meet the needs of our DE programs, we were faced with the challenge of applying them to our face-to-face programs. In the early stages of the face-to-face provision, the procedures used for distance learning programs were adopted. However, as the face-to-face provision has grown (in 2010/2011 there are more than four thousand full-time students with an average age of 22, compared to the six and a half thousand full-time equivalent distance students with an average age of 34), the applicable protocols have been amended to suit the needs of these programs. Most of these changes have been made to reflect the fact that the majority of teaching in the face-to-face programs is undertaken by members of faculty employed on a full-time basis. In addition, the majority of programs have been developed from their distance learning counterparts, which have already gone through a rigorous approval process. This has meant that the QA procedures applicable to the face-to-face programs have been adapted to take a "lighter touch" with regard to monitoring and approval. For example, the marking of continuous assessment for face-to-face courses is not subject to the same level of oversight and standardization. This is partly due to the smaller class sizes for face-to-face courses (reducing the number of individual markers) and partly due to the fact the continuous assessment for full-time courses is marked by full-time members of faculty.

This adaptation has, however, not been without its challenges. In several significant areas the University is now running two parallel QA systems that make reference to different key performance indicators. While application and admissions data are closely scrutinized for face-to-face programs in order to allow us to make comparisons with our competitors, distance learning programs have an open entry policy, and, in this case, application data is mainly scrutinized to assist with marketing plans. Student evaluation of teaching is undertaken on a sampling basis for distance learning but is universal for face-to-face courses. For face-to-face programs, the university undertakes a comprehensive graduate employment survey. However, for distance learning programs, this is not so critical, as the majority of distance students are already in employment when they commence their studies. These differences can cause some confusion, especially as many of the faculty members are involved in operating both distance and face-to-face programs. As a result, there is occasional pressure to amend the distance learning systems to match those for the face-to-face programs.

Programs in Mainland China

In the run-up to 1997, when Hong Kong was returned to the People's Republic of China, every organization in Hong Kong realized the importance of linkages with the Mainland. Hong Kong's status as a special administrative region of China confers no special privileges on its educational institutions, so in 1997, the OUHK began to launch certain programs at the postgraduate level in partnership with Mainland universities and professional organizations.

While the Mainland market is vast and has apparently endless growth potential, maintaining the quality of educational programs offered there is not without its challenges. The OUHK has worked hard over the years to establish the integrity of its assessment processes in Hong Kong as a fundamental indicator of the quality of the degrees it awards. We go to great lengths to ensure that we form partnerships only with high-quality and reliable Mainland partners. Even so, in order to ensure the security of the examination process in its Mainland ventures, the University found it necessary to send its own staff to oversee the examination operation. These additional QA mechanisms add significantly to the costs associated with running the operation, but are deemed essential to its long-term sustainability.

Introduction of e-learning

Hong Kong is a technologically advanced society. It has a highly developed technological infrastructure with a home internet usage rate of 73 percent in 2009 and one of the highest mobile phone ownership and usage rates in the world, with one thousand seven hundred and twelve subscribers per thousand population in the same year[1].

---

[1] http://www.censtatd.gov.hk/home/index.jsp

The OUHK launched its Online Learning Environment (OLE) in 1999 and now all courses have an online presence. Individual courses use the OLE in different ways; this can vary from the use of online discussion boards and electronic submission of assignments to the provision of specially developed electronic course materials.

The University has now launched the first in a series of fully online programs, where not only are the course materials provided electronically but real time tutorials are also delivered via streaming technology. In theory, students from anywhere in the world can register for these programs. In an attempt to globalize its electronic programs, the University has entered into partnership agreements with various overseas organizations.

While the introduction of online instruction has significantly improved the opportunities for tutors to provide individualized and general learner support, enhanced student-to-student communication and facilitated monitoring of the teaching and learning process by the Course Coordinator and the external examiner, it has also raised staff development and workload issues for both part-time tutors and full-time academic staff. The University now provides specialist online training for tutors to enable them to carry out their e-tutoring role more effectively.

Collaborative Programs

Over recent years the University has entered into a number of collaborative arrangements: in the Mainland, for online programs, and with various local institutions which act as feeders for our full-time programs. This has prompted us to develop procedures governing the quality and standards of collaborative enterprises. These procedures cover such topics as the role of collaborative programs, selecting a partner organization, specifications for written agreements with partner organizations, roles and responsibilities with regard to academic standards, and financial and resource arrangements. The bottom line for any collaborative arrangement is that if the award granted is in its name, then the OUHK is responsible for the academic standards of that award.

LESSONS AND CONCLUSIONS

Successful oper. and distance learning institutions have built their reputations on the reliability and consistency of their course development and delivery systems, upon the integrity of their assessment systems, and the recognition of the credits they award (Hope 2005: 152).

This case study has demonstrated that there is more to achieving a quality culture than the establishment of QA mechanisms. The original QA systems that were operated by the OUHK at its outset were imported and, as such, they neither received wholesale support from faculty members, nor were they exactly fit for purpose. They did, however, enable the institution to get off the ground quickly and meet external validation and accreditation requirements. This required a top-down, somewhat heavy-handed approach, and the initial reaction of staff was to comply rather than embrace their responsibility for quality. Over the last 21 years of operation, the systems have been refined, simplified, and adapted to meet changing needs and operational requirements and to modify the culture to make quality "everyone's business".

To achieve a quality culture, we have learned that responsibility for quality should be situated as near as possible to the "sharp end" of the process being evaluated. If, in its enthusiasm for quality, an institution establishes QA procedures with too many levels of oversight, faculty members will feel distrusted and devalued as key participants in the process. Nevertheless, quality must continue to be championed from the top.

The self-financing nature of the University can provide challenges with regard to maintaining financial stability, while at the same time keeping student fees affordable and providing a high-quality educational experience. Quality incurs costs and yet there is little tangible reward for expenditure on QA. Nevertheless the quality imperative is real, since open and distance learning is, by its very nature, more open to public scrutiny than its face-to-face counterpart.

As the University moves into a new era in which it embraces both distance and face-to-face learning and caters to the needs of both part-time adult and full-time 18–21 year old students, it has been forced to further adapt its QA systems to reflect the critical differences between the two operating modes and ensure that in all aspects of its provision it can continue to meet the requirements of the external accreditation authority and safeguard the reputation of the institution.

REFERENCES

Butcher, R.E., and Hope, A. 2006. "Embracing change: Quality assurance at the Open University of Hong Kong," in B.N. Koul, and A. Kanwar (eds), *Towards a Culture of Quality*, pp. 113–124. Vancouver: Commonwealth of Learning.

Hope, A. 2005. "Quality matters: Strategies for ensuring sustainable quality in the implementation of ODL," in A. Hope, and P. Guiton (eds), *Strategies for Sustainable Open and Distance Learning*, pp. 131–155. London: Routledge.

Hope, A. and Butcher, R.E. 2005. "History of Distance Education in Hong Kong," *Quarterly Review of Distance Education*, 6 (3): 207–215.

Kiloh, G. 1999. "The First Registration," in L. Chow (ed.), *Learning for all: The First Ten Years of the Open University of Hong Kong*, pp. 50–51. Hong Kong: Open University of Hong Kong Press.

Government of Hong Kong 1989. The Open Learning Institute of Hong Kong Ordinance, (amended in 1997 to become the Open University of Hong Kong Ordinance).

The Open University of Hong Kong. 2008. *Institutional Review 2008*. Hong Kong: OUHK in-house publication.

―――. 2010. *OUHK Quality Assurance Handbook*. Hong Kong: OUHK in-house publication.

Robertshaw, M. 1997. "Developing quality systems in the fast lane: The Open University of Hong Kong," in A. Tait, (ed.), *Quality Assurance in Higher Education: Selected Case Studies*, pp. 67–76. Vancouver: Commonwealth of Learning.

S. Korea's Hanyang Cyber University

4

Yeonwook Im

INTRODUCTION

Korea has been at the forefront of systematically developing cyber universities to provide opportunities for lifelong learning to people in Korea and other parts of the world (Lee, Leppisarri, and Im 2009). By offering asynchronous e-learning programs, Korean cyber universities have greatly expanded opportunities for higher education to a diverse population. Both the curriculum and general academic management are provided using online systems that operate without the limitations of time and space (Im and Bautista 2009). Unlike traditional universities that use e-learning to supplement classroom instruction, the cyber universities offer bachelor's or master's programs entirely online. Any adult student with a high school degree can enter a cyber university and study at any time and from anywhere. The cyber universities have played an important role in broadening access to higher education for those who have not received bachelor's degrees, those who want to change their career path, and those who want to enhance their knowledge and skills to realize a better future for themselves and their communities.

To improve the quality of education provided by the cyber universities and ensure accountability for public funds, the Korean government has introduced a number of quality assurance (QA) measures, including the initial accreditation, ongoing regular evaluation, new degree or certificate program accreditation, and regulation of student admissions. While the cyber universities need to have a more flexible academic system to facilitate lifelong education, appropriate QA policies are important. Considering that the cyber universities educate students who are remotely situated from their instructors, how well the students learn in the process of gaining their degrees should be measured regularly. Both the cyber universities and the Ministry of Education, Science and Technology (MEST) in Korea focus on QA to enhance and fortify educational policy.

This chapter presents the QA mechanisms at the Hanyang Cyber University (HYCU), one of the leading cyber universities in Korea. HYCU, established in 2002 as one of the first cyber universities in Korea, currently has the largest student enrollment among the cyber universities in Korea, and became the first to be granted permission to offer graduate programs by MEST.

The next section provides an overview of QA issues facing the cyber universities in Korea, followed by an analysis of HYCU. The chapter will conclude by addressing future issues in QA in e-learning.

CYBER UNIVERSITIES IN KOREA

Profile of Cyber Universities

The concept of cyber universities was first introduced in Korea by the Presidential Committee for Educational Reform in 1996, and piloted between 1998 and 2000. The main purpose for establishing cyber universities was to expand opportunities for higher education, especially for people who were unable to attend traditional universities. This goal was in keeping with efforts to enhance the quality of adults' lives by providing sustainable education. Cyber universities were also considered the means to train workers for a knowledge-based society.

During the pilot period, MEST, in collaboration with higher education institutions and educational technologists, developed a set of quality criteria for establishing and monitoring a cyber university, particularly in terms of financial status, academic purposes, preparation of equipment, human resources, instructional design, and interactions. In 2001, nine cyber universities were approved by MEST and began to accept students. Since then, the number of cyber universities has increased to 20, with the total number of students enrolled reaching more than ninety thousand in 2011 (MEST 2011). The MEST quality evaluation criteria have also been updated to incorporate changes in e-learning environments and adult learners' needs.

With the advancement of e-learning technologies and wide acceptance of e-learning came an increase in the enrolment of

Figure 4.1 Cyber universities' student distribution by age

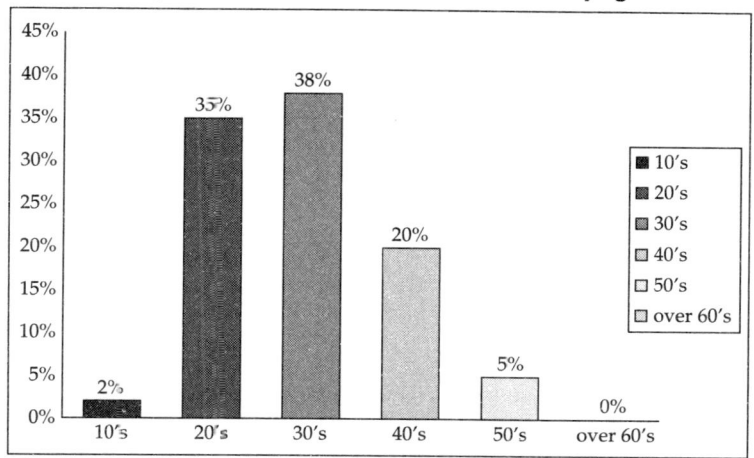

Source: Adapted from MEST and KERIS (2008).

students from diverse background in the cyber universities. Figures 4.1 and 4.2 show the diversity of students in terms of age and academic background.

Figure 4.2 Cyber universities' student distribution by academic background

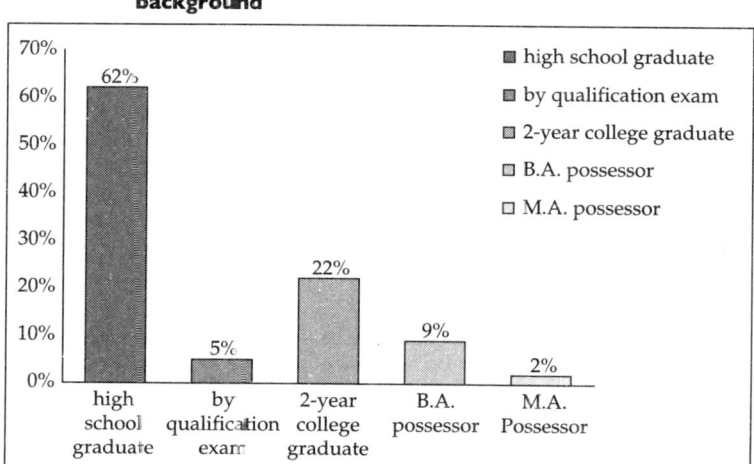

Source: Adapted from MEST and KERIS (2008).

Table 4.1   Rate of annual enrollment

Year	Annual admission quota openings	Actual annual enrollment	Number of cyber universities
2001–2002	16,250	9.920	15
2002–2003	20,600	10,987	16
2003–2004	22,600	10,459	17
2004–2005	23,550	14,620	17
2005–2006	23,550	18,138	17
2006–2007	23,550	16,787	17
2007–2008	26,200	21,001	17
2008–2009	27,960	22,814	18
2009–2010	29,400	23,975	19

Source: MEST and KERIS (2011).

Over the past ten years, the admission quota for new students has steadily increased every year (see Table 4.1). MEST monitors and adjusts the admission quota every year as a method of quality control. Universities that would like to go beyond their admission quota need to seek approval from MEST.

QA in Cyber Universities

With the rapid growth of cyber universities, it has become crucial to assure their educational quality in order to maintain the value of the degrees awarded and the standard of Korean higher education in general. MEST has devised a two-part system for QA: initial institutional accreditation and periodic academic monitoring. In addition, since 2008, MEST has approved the establishment of new graduate programs at the cyber universities.

Initial institutional accreditation: A cyber university is a higher education institution utilizing information and communications technology. Korean cyber universities are all private institutions, based on the "law of private schools." Individuals who wish to establish a cyber university should provide the necessary capital and follow the procedures for accreditation. As prescribed by MEST, a cyber university should fulfill the following educational purposes:

- Expansion of opportunities for higher education for adult students;

- Guarantee of the people's right to learn through diverse educational services to meet their needs;
- Expansion of educational services for workers, handicapped people, and other marginalized people;
- Offering of high-quality education using various technologies and educational methodologies;
- Contribution to the strengthening of national competence by cultivating experts, and through the reeducation of adults; and
- Provision of good quality education at lower cost.

The accreditation procedure is as follows. First, an educational foundation that wishes to establish a cyber university forms a Cyber University Establishment Committee. The Committee submits an establishment plan to MEST. MEST then requests the Cyber University Establishment Evaluation Council to review the establishment plan. Based on the results of the review from the Evaluation Council, MEST makes a decision to approve or disapprove the establishment scheme. Once the scheme is approved, the Cyber University Establishment Committee submits an authorization application form to MEST. Again, based on the review results from the Evaluation Council, MEST makes a final decision whether or not to approve the establishment of the cyber university. Figure 4.3 illustrates the accreditation procedure.

The evaluation by the Evaluation Council includes quantitative and qualitative analysis. For the quantitative analysis, the applicant's compliance with the establishment criteria, clauses 5 to 7 of the Regulation of Cyber University's Establishment and Operation,

Figure 4.3: Initial institutional accreditation procedure

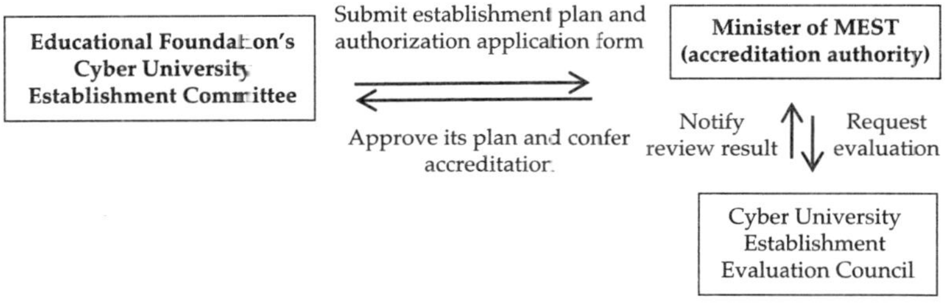

Source: Adapted from MEST (2010).

Table 4.2　Evaluation criteria for qualitative analysis

Requirements	• Appropriateness of mission statement and name of the cyber university • Rationale of establishment and development plan • Validity of the analysis of target learners and the demand for the educational service • Plan for the establishment of school buildings • Plan for faculty and other human resources development • Acquirement of basic properties • Validity of financial and management plan
Facilities for Education and Research	• Appropriateness of infrastructure, such as the server and communication systems • Stability and appropriateness of the operation platform and software • Having a digital library
Academic Administration	• Appropriateness of the curriculum • QA measures for academic operations • Appropriateness of admission policies • Enactment of school regulations and the Charter
Human Resources and Organization	• Appropriateness of human resources such as administrative staff, content development team, and system operation team
Management and QA of Cyber Educational Programs	• Validity and concreteness of content development system • Appropriateness of content development procedures • Appropriateness of content QA • Appropriateness of content maintenance system

Source: Adapted from MEST (2010).

is rated Pass or Fail. The evaluation criteria for the qualitative analysis are listed in Table 4.2.

Documents for submission include the following:

- Cyber University establishment plan
- Facilities and faculty development plan
- Prospectus for securing school buildings
- Prospectus for securing faculty
- Prospectus for securing teaching assistants and tutors
- Prospectus for raising funds
- Budget of the educational foundation (for the last two years)

- Prospectus for managing University finances (for the last four years)
- Prospectus for securing permanent property for-profit
- General chart for securing educational facilities and equipment
- Prospectus for securing servers and communication infrastructure
- Prospectus for organizing the distance education system
- Prospectus for managing facilities and equipment of media studios
- Prospectus for managing academic operations
- Prospectus for managing distance education programs and QA

Periodic academic monitoring: Once a cyber university is established, it is subject to regular monitoring and evaluation by MEST. The basic evaluation criteria are similar to those applied in the initial accreditation. The first comprehensive evaluation of cyber universities was undertaken in 2007. The evaluation included six areas: educational planning, teaching and learning, human resources, material resources, management and administration, and educational outcomes. These evaluation areas were derived from the educational system model for cyber universities (see Table 4.3). Table 4.4 shows detailed evaluation criteria and scoring method. All cyber universities are required to undergo

Table 4.3 Educational system model of cyber universities

	Educational objectives	Teaching and learning system	Support system	
Input	Educational plan (philosophy, purpose, objectives, curriculum)		Human resources	Material resources
Process		Teaching and learning (course design, content development, course implementation, learning support)	Management and administration	
Output	Educational outcomes			

Source: MEST (2007).

Table 4.4   Monitoring criteria and scoring method

Evaluation areas	Evaluation section	Point distribution Ratio (%)	Point distribution Score (point)	Area score (point)
1. Educational planning	1–1 objectives	3	15	45
	1–2 curriculum	6	30	(9%)
2. Teaching and learning	2–1 course design	6	30	
	2–2 content development	6	30	155
	2–3 course implementation	14	70	(31%)
	2–4 teaching and learning evaluation	5	25	
3. Human resources	3–1 student	4	20	100
	3–2 professor	10	50	(20%)
	3–3 staff	6	30	
4. Material resources	4–1 facilities	3	15	75
	4–2 equipment	4	20	(15%)
	4–3 system	8	40	
5. Management and administration	5–1 management	8	40	80
	5–2 administration	8	40	(16%)
6. Educational outcomes	6–1 satisfaction	5	25	45
	6–2 social recognition	4	20	(9%)
Total		100	500	500 (100%)

Source: MEST (2007).

the periodic monitoring process. The evaluation results affect the admission quota for new students as well as research funding from MEST. The cyber universities are also required to upload the annual QA results on their homepage.

Accreditation for graduate programs: Before 2008, all cyber universities were regulated under the Law of Lifelong Education as lifelong education institutions. In 2008, among 18 cyber universities, the 12 cyber universities that passed the MEST evaluation were placed under the Law of Higher Education and permitted to change their status to a regular higher education institution, while the six that did not pass the evaluation remained under the Law of Lifelong Education. The same laws and regulations thus now govern both cyber and conventional universities. As a result, the cyber universities have been able to create graduate programs. This recognition of cyber universities as higher education institutions has considerably improved their social and legal status in society. The cyber universities that come under the Law of Higher Education must conduct an annual self-evaluation, report the results to MEST, and publish the results on their homepages, just as the conventional universities do. But unlike the conventional universities, the cyber universities also undergo the government's periodic academic monitoring.

In 2010, HYCU became the first cyber university to receive accreditation for its graduate school programs. Three more cyber universities were similarly accredited to offer graduate programs in 2011.

QUALITY ASSURANCE AT HANYANG CYBER UNIVERSITY

Overview

Programs

HYCU was founded in March 2002 by one of the leading private conventional universities in Korea, the Hanyang University. It began with approximately nine hundred and fifty students in five departments By 2011, HYCU had created 15 undergraduate programs in five schools and eight graduate programs in three schools. As of 2011 it serves nearly twelve and a half thousand students, about five hundred of which are at the graduate-level. The majority of HYCU students are in their 20s and 30s. HYCU has 74 full-time faculty, one hundred and fifty nine part-time faculty, and 86 tutors. It also has 49 full-time and 22 part-time administration staff. Tables 4.5 and 4.6 list HYCU's academic programs as of July 2011.

Teaching and Learning

All courses at HYCU are offered on its self-developed Learning Management System (LMS) that is updated on a regular basis, based on feedback from faculty and students. The LMS includes discussion rooms, chat rooms, and Q&A rooms to promote interaction. Students are evaluated through assignments and real-time examinations. Figure 4.4 shows how students undertake a course and how learner–learner and learner–teacher interactions occur.

Table 4.5   HYCU's undergraduate programs

School	Number of students	Programs
Engineering	555	– Computer Engineering
Sciences	486	– Information and Communication Engineering
Humanities/	440	– Educational Technology
Education	1055	– English
	755	– Child Studies and Education
	1284	– Counseling Psychology
	67	– Japanese Language
Social Sciences	1390	– Real Estate
	1603	– Social Welfare
	292	– Silver Industry Management
	61	– Health Administration
Business	1992	– (School of) Business Administration
	395	– Advertising & Media
	426	– Hospitality and Tourism Management
Design	1246	– (School of) Design

Source: The author has created this table.

Table 4.6   HYCU's graduate programs

School	Number of students	Programs
Graduate School	116	– MBA
of Business	44	– GreenTech MBA
Administration	33	– Hospitality and Tourism MBA
	44	– Information Technology MBA
	39	– Media and Entertainment MBA
Graduate School of	44	– Child and Family Studies
Human Services	76	– Counseling Psychology
Graduate School of Real	55	– Real Estate
Estate		

Source: The author created this table.

HYCU offered about three hundred courses in 2010, supported by nearly 200 professors and over 50 tutors each semester.

As Harasim (1993) notes, online discussion is a useful methodology for developing higher-order knowledge through social interaction and dialogue. In order to reduce dropout, it is important to maximize interaction among students and between professors and students, as this enhances the sense of a collaborative

Figure 4.4 How to take a course at HYCU

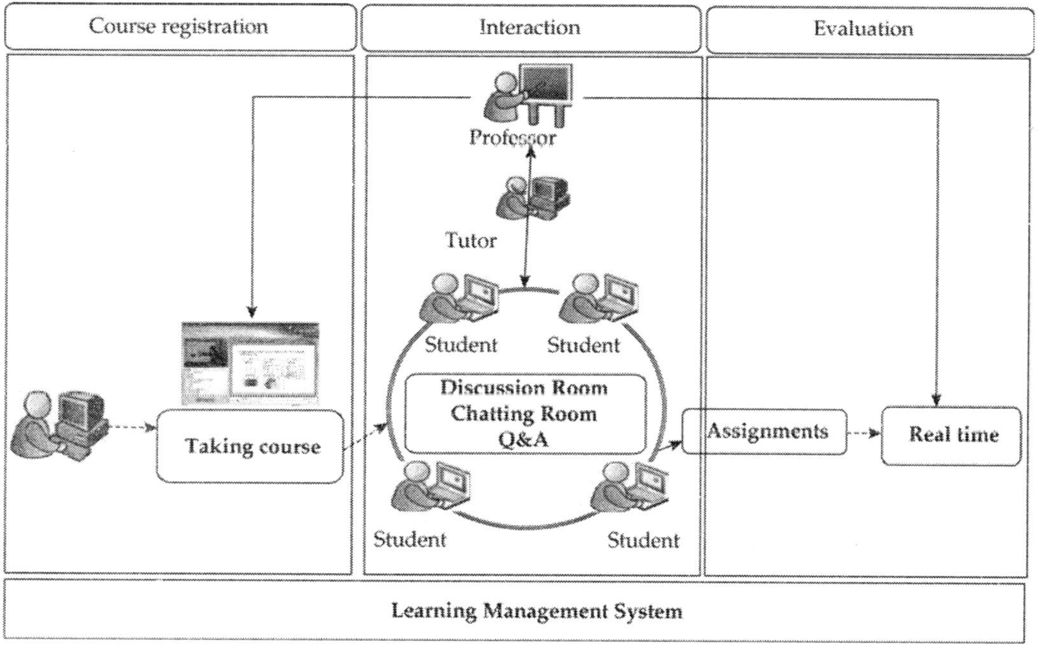

| Course registration | Interaction | Evaluation |

Taking course

Professor

Tutor

Student
Student
Student
Student
Student

Discussion Room
Chatting Room
Q&A

Assignments

Real time

Learning Management System

Source: HYCU (2011).

community (Im 2007). In this regard, HYCU supports both synchronous and asynchronous online discussion, as well as private one-to-one consulting between a professor and a student in a separate room on the LMS. In addition, video conference seminars are offered for graduate programs to help students deepen their knowledge through real-time discussions. Each synchronous seminar class has up to 25 students.

Student Support

In general there is a high degree of transactional distance between Korean students and their professors. As a result, students do not usually engage in consultations with their professors. However, the online tutor connects professors and students and assists both parties to maximize student learning. Every department at HYCU has one or two full-time tutors and several part-time tutors to provide fast and reliable tutorial support. Full-time tutors can monitor all of the courses in their departments, while part-time tutors have access to only the course assigned to them. Most of the tutors have master's or higher degrees.

Tutor responsibilities include guiding students regarding the academic schedule, answering students' academic or technical questions, sending e-mails or text messages to students who are behind, counseling, providing guidance on homework and/or the e-learning method, supporting professors with class management, monitoring examinations, reviewing content, and providing support for disabled students.

The majority of new or transfer students are full-time employees who stopped their studies for career purposes. This unique situation causes them difficulties in studying online, preparing for tests, completing assignments, and cultivating a sense of belonging. All of these can contribute to student dropout. HYCU has implemented a "Mentoring Program" to help new students who may experience these problems to adjust to cyber university life. The mentors are experienced senior students. The mentor–mentee groups are selected from the same department and they work together for one semester. The mentors play an important role in providing advice on online study methods or school life in general. The mentees share information and help one another to maintain their learning motivation.

Through an agreement between the Hanyang University and HYCU, HYCU students can acquire credits from the Hanyang University and vice versa. Also, HYCU students can access any material from the Hanyang University Library. Both HYCU and the Hanyang University students are provided with the Library Wireless Application Protocol (WAP) service. A cell phone or laptop with wireless network capability can be used to search materials. HYCU students can also use the Hanyang University's hospital services at half the usual price.

Promoting continuing education and research: The Center for Continuing Education opened in 2002 to offer various nondegree programs such as General Studies, Foreign Languages, Information Technology, Korean Studies, Korean language, and other Certificate programs. As of 2011, around one thousand students are enrolled in 2C programs. For example, the Korean language program currently offers beginner courses (levels 1 and 2) covering 30 modules over three months. A certificate of completion is given to students upon successful completion of each course.

HYCU's research institute opened as the "Sustainable Development Institute" in June 2007. It aims to provide professors with a supportive research environment. There are six research centers under the institute: Sustainability, Engineering Technology, Human Services, Industry, Business and Design. It has published five volumes of the *Journal of Sustainability Research* since 2008.

Quality Assurance Mechanism

Ensuring the quality of online education is an arduous task. As part of the Hanyang University, with its 70-year exemplary track record as a prestigious conventional university in Korea, HYCU likewise upholds policies of excellence in academic administration, faculty, content, and a student support network. To implement these policies, HYCU has adopted a number of QA measures, as now discussed.

QA for Content Development

HYCU provides various media formats and structures of content depending on course objectives and students' needs. There are

16 types of media and 13 different structures used to develop content. Media types include audio-on-demand, video-on-demand (VOD), streaming videos, Web-based training (WBT) + VOD for practice, team teaching, and WBT offline lecture. The structures include lectures engaging practitioners or specialists, lectures focusing on critical analysis, lectures adopting drills and practices, field studies, Q&A lectures, seminars, problem-based learning, and case-based learning.

The content development follows eight steps from course selection to implementation, to ensure the quality content development, as shown in Figure 4.5. A professor and an instructional designer work as a team from the first planning stage to the final editing stage. They brainstorm, select the content structures and media, develop screen drafts, and monitor the whole development process. The content development process and the quality of the courses produced are overseen by a QA committee.

HYCU develops and manages its own Learning Content Development and Management System (LCDMS) that facilitates content production. The system, which is web-based, supports communication between faculty and developers, and manages schedules and content development histories. In order to develop top-quality content, HYCU has also established several production studios and hired media specialists.

As a result of these QA measures, HYCU's content has been recognized as exceptional by the Korea Education and Research Information Service (KERIS) over the past five years.

QA for Course Implementation

HYCU's course implementation team is responsible for ensuring the quality of online classes. The team consists of administrative staff and tutors. To prepare for a new semester, faculty and tutors participate in an orientation session about the system and course management.

First, the team helps students check on their registration and develop the semester's plan and academic regulations. Secondly, the team monitors courses, checks on content uploads, manages enrollment and presence, checks on student learning status, and encourages students with difficulties. If there are students

Figure 4.5 Process of content development

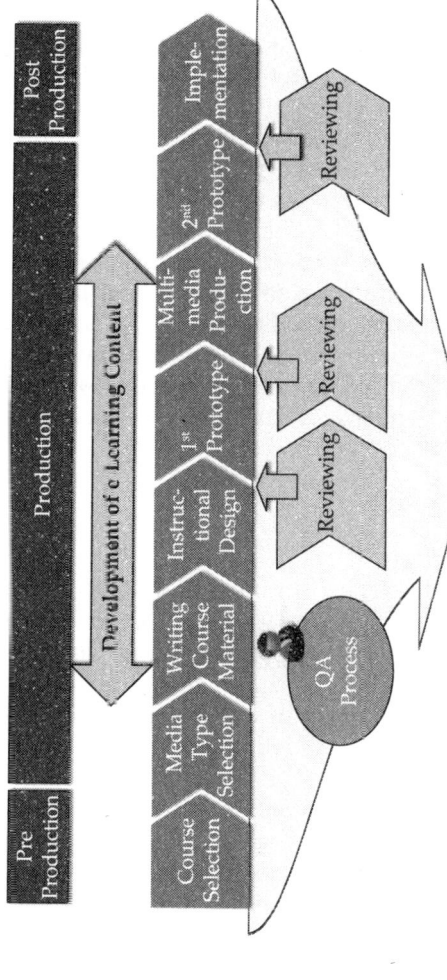

Source: HYCU (2011).

struggling with their studies, the team sends emails and/or text messages to extend support. Thirdly, the team manages examinations. In addition, the team announces the test schedule and format in advance, manages and develops test programs, conducts examinations, and announces examination scores. Lastly, the team manages grades by supporting faculty with grade inputs on the LMS, monitoring the grade viewing system and grade modification process, confirming final grades, and sending report cards to the students.

At HYCU, the notion of blended learning is embedded in every department. Offline seminars are provided regularly and offline study groups are supported. Students who are struggling with their online studies or need to be motivated are encouraged to participate in offline or online communities.

HYCU has a QA committee to monitor and improve course implementation and student satisfaction. The committee plays an active role in such issues as the review and approval of course implementation plans, course implementation related policies, and course implementation-related research and strategic planning.

QA for Tutor System

Tutors are required to be informed and have experience of e-learning and student support, on top of their subject area expertise. HYCU provides support to tutors through training programs and teaching guidelines. Before each semester starts, an orientation session and a series of workshops are held for tutors. Various seminars are also provided once a year. Tutors must meet with their professor three times per year.

HYCU has constructed an online community for tutors to support their work and help develop their professional skills. On the online community site, they access information on course implementation, post inquiries, and submit their weekly work reports. Tutors strengthen their teaching skills by sharing opinions and engaging in discussions. Tutors are evaluated every semester, and departments do not renew contracts for any tutor who scores less than 70 out of 100. On a positive note, incentives are given to outstanding tutors.

QA for Interaction

To comply with the MEST regulation that limits class size to two hundred students and ensure the quality of class interactions, any class with more than two hundred students is split into smaller groups. As a consequence, even though HYCU offered about three hundred courses in the fall semester of 2010, there were more than four hundred and fifty groups open for students.

QA for Faculty Support

The Center for Teaching and Learning, established in 2006, is an HYCU educational site to support e-teaching and e-learning activities. It provides information about workshops, seminars and orientation sessions on e-learning, content development, teaching tips, copyright, and lecture series. HYCU also provides online orientation for professors through this Center.

Lee and Im (2007) suggest that distance education universities should provide their faculty and staff with detailed services related to academic administration, student support, student aid, and educational evaluation. HYCU's Center for Teaching and Learning holds workshops, orientation programs, and seminars for faculty on a regular basis. Orientation sessions for new faculty members are held four times a year. Im et al. (2009) have found that professors believe that creating online lectures requires three times the effort in comparison to offline lectures, since more detailed preparation is needed. This indicates that workshops and other forms of support are vital for QA in faculty performance.

QA for Academic Integrity

HYCU has been concerned about how it can prevent inappropriate behavior such as cheating or plagiarism when students take real-time tests online. The real-time test system is an effective method for grading a large number of students, but presents serious QA challenges that HYCU has addressed through a series of regulations. Students are required to have a digital certificate (used for internet banking) to take a real-time test, and they have to submit an agreement that they will not cheat. The test system also detects suspicious signs and behaviors such as more than one user with the same IP address or the opening of external software programs like an instant messenger application. In this case the students are given a warning and, if the issue persists, their test systems are shut down.

QA for Mentoring Program

In the aforementioned HYCU mentoring program, mentors are experienced students and mentees are newcomers—either freshmen students or newly transferred students from other colleges. This is a peer support program, with seniors in a department helping newcomers on such issues as academic matters or school lives. The program is monitored and managed to ensure its success. Mentors file their activity report with staff at the Office of Student Affairs. All activities of mentors and mentees are recorded and they also take part in an online discussion in their special community site. The activity reports (midterm/final) submitted by mentors and their performance records for the semester are considered for the selection of outstanding mentors. Feedback from mentors and mentees is used to guide program review and improvement. Mentors can earn volunteering credits and a certificate depending on the evaluation of their performance. If their activity does not meet specific criteria, they cannot receive credits.

ISSUES AND CHALLENGES

While developing, implementing, and modifying various QA policies and practices discussed above, HYCU has faced several issues and challenges.

Mobile Learning

Mobile learning provides a new opportunity for more flexible learning. HYCU commenced mobile services for delivering learning content in March 2011. Content is delivered on such devices as smartphones and other mobile devices. Students can study via their smartphones in a manner similar to how they study using their PCs.

However, a number of issues need to be considered when implementing e-learning on mobile devices. First, it is not easy to track students' attendance, and course success is deeply rooted in a student's participation. Secondly, privacy issues are a challenge, and HYCU has experienced hacking incidents. Such issues need to be addressed for mobile learning to fulfill the aim of promoting prompt and constant interaction between professors and students via a personalized learning service.

Class Size and Tuition Fees

To enhance e-learning quality, the student–teacher ratio should be kept to a minimum. However, in Korea it is widely believed that e-learning should cost less than classroom education. Currently the tuition fees of cyber universities are one-third of those in conventional universities. Developing high-quality content and managing classes with high levels of interactions entails considerable costs. Thus, academic administration may need to plan for two kinds of online classes: mega classes with lower tuition fees, and smaller classes with higher tuition fees. Alternatively, general courses could take more students in a class where a professor offers one-way lectures, while specialized courses could have smaller student enrollment with increased interactivity.

Research and Development

Research is essential for cyber universities to provide high-quality education services. Research is needed to evaluate student satisfaction, participation, and achievement. Conducting research, utilizing the results accordingly and promoting faculty and staff to engage in research activities are important steps toward quality improvement in cyber universities.

Evaluator Competencies and e-learning QA Standards

As discussed, periodic academic monitoring of the cyber universities has been conducted by a team of external evaluators appointed by MEST. These external evaluators are often from the conventional universities, corporations, and other private and public sector institutions, and only a few have in-depth understanding of e-learning in higher education. Training programs that assist evaluators of cyber universities are needed to develop competencies for high-quality monitoring and evaluation.

In addition, the evaluation criteria and regulations stipulated by MEST are largely based on QA standards and procedures for evaluating conventional universities. There is an urgent need to identify and reflect specific features of cyber universities in the evaluation criteria.

CONCLUSION

The case of HYCU has shown that:

- Developing and delivering high-quality educational content and maximizing the potential of such human resources as in faculty, tutors, and staff are critical factors for QA in cyber education;
- QA is critical, considering the rapid growth of e-learning and the increased influence of e-learning on lifelong education;
- Research and continuous self-evaluation are important in assuring and improving the quality of cyber education; and
- Each cyber university must develop a quality culture among its members, and the government should support the university and ensure that minimum quality standards are met.

Currently, MEST is the only agency that monitors and regulates cyber universities. Eventually the role of MEST will be limited to supporting QA as each cyber university develops a QA culture and undertakes its own QA measures. It may thus be time to consider diversifying the evaluation bodies, promoting self-evaluation of cyber universities, and focusing on support and incentives for best practice.

REFERENCES

Harasim, L. 1993. *Global networks: Computers and international communication.* Cambridge, MA: MIT Press.

HYCU. 2011. "Introduction," *Powerpoint Presentation Material of Hanyang Cyber University.*

Im, Y. 2007. "A substantial study on the relationship between students' variables and dropout in Cyber University," *Journal of the Korean Association of Information Education,* 11(2): 205–220.

Im, Y. and Bautista, D. 2009. "Conceptualizing a Cyber University Model in support of effective ESD," *Asia Pacific Collaborative Education Journal,* 5(1). Available online at http://www.acecjournal.org/2009/data1/APCJ/Volum5.1/APCJ_Vol5_No1_Conceptualizing%20a%20Cyber%20University%20Model%20in%20support%20of%20Effective%20ESD.pdf and downloaded on September 4, 2012.

Im, Y., Lee, O., Chung, M. and Lee, J. 2009. *A study on the link between e-learning contents at higher education and online classes: focusing on e-learning contents running time.* Report to the Ministry of Education, Science and Technology, Seoul, Korea.

Lee, I. and Im, Y. 2007. "In search of an academic and organization model of e-Distance Universities," *Asia-Pacific Collaborative Education Journal,* 2(2). Available online at http://www.acecjournal.org/2009/data1/APCJ/Volum2.2/InSearchofanAcademic pdf and downloaded on September 4, 2012.

Lee, O., Leppisaari, I. and Im, Y. 2009. "Guidelines for national e-learning evaluation— International comparative study between Korea and Finland," *Asia Pacific Collaborative Education Journal,* 5(1). Available online at http://www.acecjournal.org/2009/data1/APCJ/Volum5.1/APCJ_Vol5_No1_Guidelines%20for%20national%20elearning%20evaluation%20International%20comparative%20study%20between%20Korea%20and%20Finland.pdf and downloaded on September 4, 2012.

Ministry of Education, Science and Technology (MEST). 2007. *2007 Handbook of Distance University.* Seoul: MEST.

Ministry of Education, Science and Technology (MEST) and Korea Education & Research Information Service (KERIS). (2008). *2008 Adapting Education to the Information Age.* Seoul: Keris.

Ministry of Education, Science, and Technology (MEST). 2010. *Annual Report of Cyber Universities.* Seoul: MEST.

Ministry of Education, Science, and Technology (MEST). 2010. *Guidelines for Establishment of Cyber Universities.* Seoul: MEST.

Ministry of Education, Science and Technology (MEST) and Korea Education & Research Information Service (KERIS). (2011). *2011 Adapting Education to the Information ge*. Seoul: Keris.

Ministry of Education, Science, and Technology (MEST). 2011, July 5. Initiatives to strengthen competitiveness of Cyber Universities. Press release report.

PART 2

Ensuring the Quality of Management Processes

Indonesia's Universitas Terbuka

5

*Sri Y.P.K. Hardini, Deetje Sunarsih,
Any Meilani, and Tian Belawati*

INTRODUCTION

An important development in this globalised era is the adoption
of quality assurance (QA) systems to ensure that the quality and
consistency of an organization's products and/or services meet
the expectations and needs of the public. In the context of higher
education, QA has been defined as "systematic management and
assessment procedures adopted by higher education institutions
and systems in order to monitor performance against objectives,
and to ensure achievement of quality outputs and quality
improvements" (Harman 2000: 1).

Quality has been set as a pillar of Indonesian National
Education, together with access, equity and autonomy (Ministry
of National Education 2002). To ensure the guardianship of quality,
the Indonesian Government has established a National Board of
Education Standardization (Badan Standar Nasional Pendidikan
or BSNP) which has developed a set of National Education
Standards (PP no. 19/2005). All educational institutions, including
universities, are required to meet the eight quality standards with
respect to learning content, competence of graduates, learning
process, human resources, teaching and learning facilities and
infrastructure, management, finance, and student assessment.
Furthermore, as part of the national commitment toward quality,
the Government has also established a National Accreditation
Board for Higher Education (Badan Akreditasi Nasional Perguruan
Tinggi or BAN-PT), an independent institution mandated to
evaluate and accredit academic programs offered by all higher
education institutions. In line with the eight quality standards
set by the BSNP, BAN-PT has developed a set of accreditation

instruments which emphasize the importance of developing the internal QA system of an institution and of meeting all the key performance indicators for each quality standard. Any higher education institution that wants to be accredited by BAN-PT fills out and submits the completed standardized instruments, including an internal QA report, to the Secretariat of BAN-PT, which conducts a desk evaluation of the submitted documents and a site visit. Based on the results of both the desk and on-site evaluations, BAN-PT may grant accreditation status. The accreditation status granted by BAN-PT is valid for three years.

BAN-PT has also developed a separate set of instruments to accredit academic programs delivered through distance education, such as those of Universitas Terbuka. The quality criteria and key performance indicators for accreditation of distance programs are equivalent to those applied to face-to-face programs. In principle, all national quality standards that are applied to face-to-face higher education are also applicable to distance education (Belawati 2010).

UNIVERSITAS TERBUKA

Indonesia first embarked on correspondence education in the 1950s and, in 1984, the Indonesian Open University or Universitas Terbuka (UT) was established. UT was given mandates to broaden access for all Indonesians to higher education and to improve the qualifications of existing teachers who had graduated from high school level teacher training programs into full degree level or bachelor level. UT was designed to use an open and distance learning system, believed to be a flexible and accessible system that allows potential students to study regardless of their demographic, economic, and geographic situations. Starting with around sixty-five thousand students, UT now offers almost one thousand courses to over six hundred thousand students, 80 percent of whom are in-service teachers and over 60 percent are female. UT has four faculties: Economics and Developmental Studies, Social and Political Sciences, Mathematics and Natural Sciences, Teacher Training and Educational Studies, and a Graduate School. Apart from the Teacher Training and Educational Studies programs, which are only for in-service schoolteachers, all other academic programs are open to any holder of a high school diploma, regardless of age and year of high school completion.

In order to maintain its openness, flexibility, and accessibility, UT consistently implements an open registration and learning system. Students may register at any time of the year and are free to take time leave during their studies without formal notification. The learning medium is selected to ensure that no Indonesians are marginalized due to a lack of access to certain technologies. Accordingly, UT's main learning medium is printed study materials, supplemented by various nonprint materials such as CD-ROMs and online materials. Learning support is provided through various media, including face-to-face, online, or radio tutorials. Students are expected to choose the learning support medium that is most suitable to their personal preference and circumstances. Assessment, however, is highly structured, through sit-in supervised examinations in certain locations designated by UT. Since 2006, UT has offered online examinations as an alternative method for those who are unable to attend the regular sit-in examination at the scheduled times. The online examinations take place in UT's 37 Regional Offices through separate registration. In order for students to graduate from their programs, they need to complete a minimum of 144–145 credit points, achieve a minimum GPA of 2.00, and pass a comprehensive examination known as the Tugas Akhir Program (TAP). This open system has proven effective in widening access to higher education and increasing participation rates in higher education.

Operating an open policy for such a large number of students within a vast country like Indonesia requires a strong management system. The Head Office is responsible for the academic and administrative policies and developing academic programs and materials, while the daily operational activities to serve the students are devolved to the Regional Offices. To provide optimal educational services to students, UT collaborates with other institutions, including local public and private universities as well as local government agencies and offices. At UT, the QA system aims to ensure the satisfaction of students and other stakeholders in terms of both academic and academic administration services.

A decision in 2001 to operate all UT activities according to standardized levels of quality has led to equal quality services for all UT students, regardless of their domicile. At that time, however, there was no available standard quality framework for open and distance learning (ODL) systems, except for a draft QA framework developed by the Association of Asian Open Universities (AAOU), of which UT is a founding member.

Since that time, UT has continuously developed and improved its internal QA system, and regularly invited national and international bodies such as BAN-PT, the International Council for Open and Distance Education (ICDE), and the International Organization for Standardisation (ISO) auditor agencies to assess and review the results of UT's commitment to quality. The acknowledgements received from such bodies reflect both national and international recognition of UT as a quality educational institution (Belawati and Zuhairi 2007).

This chapter will describe UT's QA policies and system. Specifically, it will focus on sharing the experience of UT in using the ISO 9001 approach as a tool for strengthening the implementation of its internal QA system.

QUALITY ASSURANCE AT UT

Overall QA Policies

As mentioned, based on AAOU's draft of a QA Framework, UT started to develop policies on quality applicable to its characteristics as a higher education institution which implements an ODL system. This development took around one year and involved all stakeholders affected by the policies. The policies materialized in the *Sistem Jaminan Kualitas Universitas Terbuka* (Universitas Terbuka Quality Assurance System) in 2002. The QA policies consist of nine components, comprising one hundred and seven statements of best practice as follows.

1. Policy and planning (seven statements)
   UT determines its own mission and objectives that reflect UT's academic commitments and the needs of society.
2. Human resource recruitment and development (nine statements)

The staff and personnel management system is appropriate for the education and training services provided. The UT sets out development programs that equip staff to perform their tasks effectively.

3. Management and administration (21 statements)
   The UT has clear and effective communication channels and has efficient resource management and administration systems that enable UT to achieve its objectives. The UT is financially sound and can make reliable educational provision.

4. Learners (10 statements)
   There is a system of collecting detailed information about learners and using this information to inform all aspects of policy and planning, program and course development, support services, and the overall processes of teaching–learning.

5. Program design and development (six statements)
   Programs are designed and developed with the needs of learners, employers, and society in mind; to encourage access to quality education; and set in place assessment methods appropriate to the aims and objectives of the programs.

6. Course design and development (14 statements)
   The course syllabus and content is well-researched. The course materials have appropriate objectives and outcomes, content, approaches to teaching and learning as well as to assessment presented clearly. There is an identified process of development and review of courses.

7. Learning support (18 statements)
   Learners are supported by the provision of a range of opportunities for real two-way communication through the use of various forms of technology for tutoring at a distance; contact tutoring, assignment tutoring, counseling, and the stimulation of peer support structures. The needs of learners for physical facilities and study resources and their ability to access these are also taken into account.

8. Assessment of student learning (15 statements)
   Assessment as an essential feature of the teaching and learning process is properly managed, and reflects external standards.

9. Media for learning (seven statements)
    The selection and application of media reflect the teaching
    and learning needs in a course and are the most appropriate.
    In particular, the choice of media is based on knowledge
    of the learners' and educators' backgrounds and abilities,
    the requirements of the content, learners' access to the
    associated technology, the pedagogical design for the
    course, and the limitations of the media.

With these statements of best practice, UT developed a set
of QA manuals to guide the implementation of the standard
operating procedures and attain the set quality standards. The QA
manuals also emphasize the interrelationship among processes,
so that everyone understands that their work affects the work of
others: every step of the work is a part of a cycle. For example,
nonconformity in one process (e.g., delays in developing learning
materials) will affect the next process (a delay in printing and
distributing learning materials to students). This nonconformity
will therefore significantly affect the quality of service to students.

Once the QA policies were in place, in 2002 UT conducted a
series of self-assessment exercises, where each statement was
critically evaluated against the performance at each management level within the working units (i.e., Study Program, Faculty,
Instititute, Regional Office, etc.). The self-assessment results
(using a 1–5 scale) at each management level were integrated with
the higher levels, resulting in an agreed standard of quality for the
whole university for each quality statement. The exercise led UT
to the realization of the need to improve many aspects of quality.
The next step was to set priorities for quality improvement.

The first priority was given to developing the aforementioned
QA manuals as working guides for all processes. This involved
around two hundred staff representing all units at UT, and took
about two years to complete. Table 5.1 shows the components of
the QA policies and the number of QA manuals in each component.
As revealed in Table 5.1, UT has two hundred and fifteen QA
manuals, which specifically outline the processes in all of UT's
programs and activities. The manuals tend to heavily focus on
the flowchart of a process, and contain steps, person in charge,
timing, and quality indicators for outputs of each activity. All staff
are expected to understand the QA processes and consistently
implement them to ensure the achievement of outputs at the

Table 5.1 QA components and number of QA manuals at UT

No.	QA component	Number of QA manuals
1	Policy and planning	11
2	Human resource provision and development	30
3	Management and administration	45
4	Learners	46
5	Program design and development	9
6	Course design and development	24
7	Learner support	20
8	Learner evaluation	23
9	Media for learning	7
	Total	**215**

Source: Sistem Jaminan Kualitas UT (2002).

defined quality standards. It is noted that, regardless of the continuous efforts to maximize effectiveness, the Manuals still lack measurable goals for each of the processes. Another round of revisions is therefore already underway. The belief in continuous improvement and commitment is encapsulated in the motto that *"we write what we do, we do what we write, we check, we act, and we improve it continually."*

In line with internal efforts to establish and implement a rigorous QA system, UT seeks external validation to ensure that all efforts have indeed resulted in better quality performance. Academically, UT works with Indonesia's BAN-PT to gain accreditation for all its academic programs. Operationally, UT invites ICDE to review its quality and efforts. ICDE's quality review focuses on ensuring that institutional policy with respect to openness and flexibility, commitment to educational values, good customer relationships, and business practices are all well guarded. In addition, UT also decided to use the ISO system to check the quality and consistency of its management processes as a complementary validation of QA efforts. The following section discusses the experience of using this ISO approach as part of its QA System at UT.

Certification with ISO 9001 Standards

As an open university, UT basically operates like a business entity. The ISO 9001 for Quality Management System is considered an appropriate tool to ensure the quality of its management processes.

Using the PDCA (Plan–Do–Check–Act) approach from ISO 9001, UT is required to continuously evaluate the system and procedures both individually and as a whole in terms of assumptions used, efficiency of practices, and whether the QA manuals remain relevant. The key to effective QA is performing the evaluation openly and honestly, both by internal and external auditors. An internal audit occurs each semester, undertaken by internal auditors who have been certified by ISO 9001 training agencies. Results of the internal audit are used for continuous improvement and as tools to avoid any recurrence of faulty processes. Similarly, an external audit is performed every semester by invited external auditors from ISO auditing agencies. The outcomes of external audits affect the status of extension as well as possible termination of the ISO certification.

The adoption of the ISO system helps UT to implement the QA manuals and thus ensure the quality of all processes and outputs. As a large organization with almost eighteen hundred staff spread across head office and 37 regional offices throughout the country, it is not easy to monitor and evaluate whether all the standardized procedures formulated within the QA manuals are being implemented consistently. UT has confidence in its internal monitoring system, but believes that an external "hand" can help verify a consistent commitment toward quality. Therefore, the aim is not just to obtain the ISO certificate, but rather to help maintain the commitment of all staff within UT to consistently follow the QA manuals. This sustains and enhances the quality of all processes and outputs of the university. Based on the experience of the past five years of using the ISO system, a positive impact on staff and university performance has been observed, resulting in a significant decrease in students' complaints and problems. It is of course a "bonus" that having ISO certificates also helps to strengthen public confidence in UT.

The use of the ISO approach started in 2005, when UT sought to obtain ISO 9001 Quality Management certification for the learning materials distribution performed by the Center for Learning Material Distribution Services (Puslaba). Puslaba was chosen because of the challenging nature of the Center's task in following tight schedules to deliver learning materials to students. Any holdup in the delivery would delay the students' independent learning process, which in turn would negatively influence their chance to master the materials before the examination. Puslaba earned the ISO 9001 certificate for Learning Materials Services in 2006. As a follow-up to the process of certification, Puslaba revised its QA manuals, making them more precise and comprehensive. Puslaba also developed an umbrella QA manual for all processes in learning materials distribution. In total, to support the attainment of the ISO 9001 certificate, Puslaba applies nine QA Manuals and 13 supporting working procedures.

In 2005, UT decided to obtain ISO 9001 certification for the development process of learning and examination materials and for student services at UT regional offices. The preparation process led to revisions of relevant QA manuals and recognition of the need to develop working procedures. UT has continued to use ISO 9001 certification to improve the quality of academic administration services, with the number of QA manuals used in various management aspects shown in Table 5.2.

As shown in Table 5.2, the attainment of ISO certificates for student/learning services at regional office level was not achieved at once, but gradually according to the readiness of each individual regional office. However, upon completion, the 25 regional offices were ther integrated and the certification was merged into one ISO 9001 certificate, as they had achieved the same level of quality and working culture. In the near future, an audit will integrate all 37 regional offices into one certification.

Table 5.2   List of ISO 9001:2008 certificates

Quality management aspect	Year of ISO certificate attainment	Remark	Number of QA manuals
Learning Material Delivery Services	2006	Mostly covering activities in the Center for Learning Materials Delivery	9
Development of Learning and Examination Materials	2007	Covering activities in the faculties, Multimedia Production Center, and Examination Center	29
	2009	Integrated with ISO for Learning Materials Delivery Services	34

(Table 5.2 Continued)

(*Table 5.2 Continued*)

Quality management aspect	Year of ISO certificate attainment	Remark	Number of QA manuals
Student/Learning Services at individual Regional Offices	2007	Covering activities at 11 Regional Offices	25
	2008	Covering activities at 12 other Regional Offices	
	2009	Covering activities at 8 other Regional Offices	
	2010	Covering activities at 5 other Regional Offices	
	2011	Covering activities at the last Regional Office	
Academic Administration Services	2008	Mostly for activities in the Bureau of Academic Administration, Planning and Monitoring (BAAPM)	6
Promotion and Cooperation	2008	Mostly for activities in BAAPM	6
Student/Learning Services at 25 Regional Offices	2010	Merging of previous ISO Certificates for 25 Regional offices into 1 ISO Certificate	25

Source: Pusat Jaminan Kualitas UT (2011).

Similarly, in 2009 the audit of learning material delivery services was integrated with the audit for learning and examination materials development, enabling the certification to be merged into one ISO 9001 certificate. UT also integrated the audit of academic administration into the audit of student/ learning services at regional offices, thus avoiding extension of ISO certification for academic administration alone.

UT plans to eventually maintain only two ISO certificates, one for operational aspects, covering what are now under the ISO for student/learning services at regional offices and some additional aspects at the head office, and another covering all academic activities. The integration of different ISO aspects will enhance the efficiency of the system without decreasing its comprehensiveness.

CHALLENGES AND LESSONS LEARNED

Need for Ensuring Quality of Academic Content

The implementation of ISO 9001 has effectively improved the quality of working processes in all units at UT. However, ISO 9001 is not concerned with content, and therefore cannot be used to ensure the academic quality of learning and examination materials. Recognizing that content is a key aspect of quality, UT places the academic quality of the materials at the core of its QA activities, and continuously revises the learning and examination materials through regular reviews by both internal and external experts in related fields. With a large number of courses (approximately one thousand) currently on offer, revising the materials is not an easy task. The policy is that the materials for every course have to be revised once it reaches its fifth year of use, which leads UT to revise around two hundred courses every year. Assessment of the academic quality of the learning and examination materials is included in BAN-PT's program audit. All study programs at UT were reaccredited by BAN-PT in 2011.

Balancing Standardized ISO Procedures and Humanist Working Culture

The ISO 9001 approach has helped UT maintain high-quality and requires the documentation of all activities, so that processes can be easily evaluated and improved accordingly. However, implementation is not without challenges. It has not been easy to change the existing work culture into a "write what you do, and do what you write" culture, as required by ISO 9001. One example of this challenge is the requirement to produce minutes for each conducted meeting, which is sometimes forgotten. Even when minutes are recorded, a component often missing is verification of decisions made during the meeting. Another difficulty is the internalization of working procedures. The replacement of a Head of unit can result in a "back to square one" condition, especially if the new head has not been informed of the ISO 9001 working procedures used in the unit. An additional challenge in having such standardized working procedures is that staff may become bored as tasks become "routine," thus decreasing the human touch in performing their work. Concerns have been expressed that staff have become like "machines", making some working processes almost unbearable. A team has been established to look into these issues and see what the university should do to balance the application of standardized procedures and the maintenance of a humanist working environment.

Building Personnel and Internal Audit Capacity

UT needs people with certain qualifications and competencies to effectively carry out required tasks. However, it has not been easy to appoint the right people to the right positions while at the same time recruiting local people for each regional office. The lack of suitable local people to head UT's regional offices can sometimes be problematic. "Parachuting" personnel from the UT head office to the regional offices needs special attention in order to maintain a harmonious environment in the regional offices. To address this issue and increase personnel competencies, UT operates continuous staff training and has increased internal audit capacity by sending personnel to gain internal auditor certification. By doing this, UT can conduct more effective internal quality audits prior to the external audit process by the ISO agencies.

CONCLUSION

It has been a long journey since UT first developed its QA system based on the AAOU draft of a quality framework in 2001. Over the past 11 years, UT has initiated many new developments in both QA frameworks and standards, and AAOU has revised its QA framework. At the national level, the Ministry of National Education in Indonesia has launched a new QA system for higher education (Sistem Penjaminan Mutu Perguruan Tinggi or SPM-PT), which will soon be in effect. In addition, Indonesia's BAN-PT has also developed a special accreditation instrument for ODL (UT), which contains quality elements and indicators to be addressed. Likewise, ICDE has also developed quality measures that are used to review ODL institutions and systems, on request. These recent developments and changes mean that UT needs to integrate all quality elements and indicators into a comprehensive QA framework that encompasses the QA policies of SPM-PT, BAN-PT, ICDE, and ISO 9001. The integration of the four QA systems into one comprehensive new system will allow UT to be internationally recognized while at the same time follow national standards.

REFERENCES

Belawati, T. and Zuhairi, A. 2007. "The Practice of Quality Assurance System in Open and Distance Learning: A Case Study at Universitas Terbuka," *International Review of Research in Open and Distance Learning*, 8(1). Available online at http://www.irrodl.org/index.php/irrodl/article/view/340/782 and accessed on September 4, 2012.

Belawati, T. 2010. "Quality assurance," in T. Belawati and J. Baggaley (eds), *Policy and Practice in Asian Distance Education*, pp. 49–59. New Delhi: SAGE Publications India Pvt. Ltd.

Harman, G. 2000. *Quality Assurance in Higher Education*. Bangkok: Ministry of University Affairs and UNESCO PROAP.

Ministry of National Education. 2002. *Higher Education Long Term Strategy*, 2003–2010, Part I, Chapter 2.

China's Peking University School of Distance Learning for Medical Education

6

Chen Li, Shen Xinyi, Gao Shuping, and Liu Yiguang

INTRODUCTION

Distance education (DE), including online education, has achieved remarkable development in China since China initiated the ICT-supported DE pilot project in 1999. Currently, there are 68 online colleges operated by the conventional universities, and the Open University of China system (OUCS).

All higher education providers in China—both conventional and distance—are required to obtain initial accreditation from the Ministry of Education (MOE) before establishment. Once established, a conventional higher education institution should undergo an academic audit of their undergraduate teaching every five years. The audit consists of self-report, external evaluation, and self-improvement.

DE institutions do not need to follow this audit process. Instead, they are required to conduct three measures specified by the MOE:

1. *Follow standardized syllabi and examinations:* To enhance the quality of DE, in 2005 the MOE set out standardized curricular syllabi and national examinations in such subject areas as *Literature, Mathematics, English* and *Introduction to Computer Application*. Distance learners who want to obtain an undergraduate diploma have to take these standardized examinations. Up to 2011, over 40 million people have taken these exams. Online exams have been completed by over 25 million people in two subjects, *English* and *Introduction to Computer Application*.

2. *Annual Reporting and Censorship (ARC):* Since 2001, OUCS and 68 online colleges have adopted the ARC system and conducted self-evaluation of the quality of their programs and services, received peer review, submitted an annual report to the MOE, and obtained feedback from the MOE. During the ARC process, Provincial Education Offices evaluate off-campus learning centers.

3. *Receive external reviews from a consortium:* A consortium, or "Third Party", composed of department heads, academic staff, and technology officers of the DE institutions, was founded in 2000 to share common problems, promote information exchange, and foster the development of DE in China. All DE providers receive periodic external reviews from this consortium.

PEKING UNIVERSITY'S SCHOOL OF DISTANCE LEARNING FOR MEDICAL EDUCATION

In 2000, Peking University established the School of Distance Learning for Medical Education (SDLME) as a subsidiary entity to provide online education in the field of medical science. SDLME adopts a corporate management system, meaning that it is operated by Beijing Bytime Science and Technology Development Co. Ltd., under the supervision of the Health Science Center of Peking University.

SDLME has 46 partners as its off-campus learning centers across 20 provinces, autonomous regions, and municipalities. These centers are accredited by the local government, and considered highly important in recruiting students, providing a supportive learning environment, and organizing cultural activities in local areas.

SDLME was the first DE provider in China to receive ISO 9001:2000 Quality Management System Certification in 2003 for its standardized management procedures, rich resources, satisfactory services and reliable technology infrastructure. Having received strong financial and faculty support from Peking University, SDLME has set the goal of "making excellent medical education accessible to all" and developed unique DE programs combining satellite, internet technologies, and multimedia. Over the past few years, it has earned a high reputation in the field of medical sciences and has been listed as one of the top quality management DE institutions by the MOE.

SDLME offers two programs: the Diploma Education Program (DEP) as its core area, and the Non-diploma Continuing Education Program (NDCEP). DEP includes three programs across several majors: one for online learners seeking a secondary vocational certificate, another for those aiming to obtain a higher vocational degree, and a third for those aiming at an undergraduate degree. It also permits credit-hour based enrolments. DEP's majors include Nursing, Applied Pharmaceutical Sciences, Health Industry Management, and Medical Information Management. Online learners enrolled in DEP study their courses using three kinds of materials: CD, online courseware, and print-based materials. Course evaluation is usually paper-based and undertaken in the learning centers, though in some cases it is completed online.

Unlike DEP, NDCEP does not target certificate- or degree-seeking students. It offers continuing education courses, using the rich medical educational resources of Peking University, with the aim of contributing to sustainable development and growth in diversified fields of the public interest. It uses its own Medical Education Information System or MEIS, which blends broadband digital satellite with high-speed internet access, to broadcast academic resources. NDCEP's database contains varied forms of educational resources such as lectures, case discussions, clinical rounds, and surgery demonstrations produced by two hundred and twenty experts and scholars from over 20 hospitals. The content, covering all disciplines in clinical medicine, provides health care institutions in China with long-term, standardized, and systematic medical training services, and is frequently used by medical service staff throughout the country. To disseminate the content, NDCEP collaborates with over one hundred health care institutions in China, and reaches as many as one hundred thousand medical service staff (Peking University School of Distance Learning for Medical Education 2011).

QUALITY ASSURANCE MECHANISMS AND PROCEDURES AT SDLME

Since its establishment, SDLME has followed the MOE guidelines listed above. It has adopted the standardized syllabi and examinations, submitted an annual report and received MOE's feedback, and regularly invited external reviewers from the "Third Party" consortium. During this process, SDLME has learned that an efficient and effective quality management system is needed to assure, and continuously improve, the quality of its programs and services. It concluded that the MOE guidelines were not enough to achieve this purpose, and that obtaining the International Organization for Standardization (ISO) certification would help develop such a system. The following subsections explain how SDLME has established its quality management system, adopting ISO's QA principles.

QA Structure

The Office of Quality Assurance is the central quality unit that monitors and analyzes the operational process of SDLME's quality management system, and supervises the quality of teaching and learning. It applies the ISO principles of QA and continuous quality improvement to guarantee effective quality management. More specifically, the Office:

- Offers staff training and internal auditor training;
- Identifies the product and its implementation process;
- Develops the school's quality policies and objectives;
- Prepares quality management system documentation;
- Carries out management reviews and internal audits; and
- Develops corrective and preventive measures.

In addition, a working group monitors and measures the quality of DE at SDLME. The vice president of Peking University is the group head, in charge of assuring the quality of distance teaching and learning at SDMLE. Group members comprise the president and vice president of SDLME, the director and vice director of the administrative affairs office, and three to five DE experts. The group reviews the appropriateness of educational planning and relevant administrative support, and regularly discusses the results of DE quality monitoring activities. Furthermore, it regularly invites external experts to check the teaching processes at SDLME. All these efforts are directed toward producing high-quality graduates.

QA duties are carried out by several offices and departments at SDLME. Table 6.1 lists the QA duties of those offices and departments.

QA Activities

The ISO 9001:2000 standard requires an organization that wishes to obtain its quality certification to stipulate and carry out an internal audit plan at least once a year.

In December 2002, SDLME started its internal audit for the first time, reviewing the services of all departments and offices, excluding extracurricular study centers. Results of this audit exercise revealed that the department chairs and office heads did not fully understand the ISO quality management system and the need to prepare documentation files. After the first audit, SDLME undertook to train its chairs and office heads regarding the importance of ISO 9000 and its detailed requirements. Results of the second internal audit in January 2003 demonstrated an improved understanding by the chairs and heads with regard to the quality system and the importance of the documentation.

The ISO approach aims to help DE institutions to develop their own quality framework and standards, evaluate the quality management system at certain intervals, put forward the need for change, and ensure their continuing suitability, adequacy, and effectiveness. It is usually carried out by an external audit team. The auditor's role is to give an objective, fair, and independent evaluation in order to help an institution improve its quality system and enable an accreditation agency to make an informed accreditation decision. In case of SDLME, the Beijing Union Limited company conducted the external audit, pointed out strengths and weaknesses in SDLME's teaching and learning processes, and made valuable suggestions for improvement. The audit team also made several valuable comments and suggestions to help SDLME improve its teaching and supporting processes. Once the external audit was completed, SDLME made improvements on suggested items and submitted a final report to seek ISO quality certification.

Table 6.1 QA duties of offices and departments

QA duty	Office of quality assurance	Office of administrative affairs	Academic departments	Office of students affairs	Technology office	Training office
Manage overall QA	◆	◇	◆	◇	◇	◇
Manage documents		◆	◇	◇	◇	◇
Manage records		◆	◇	◇	◇	◇
Assign management responsibilities	◆					
Implement learner-centered policies	◆	◇	◇	◇	◇	◇
Develop QA policies	◆	◇	◇	◇	◇	◇
Develop QA objectives and system planning	◆	◇	◇	◇	◇	◇
Review performances of all responsibilitie	◆	◆	◇	◇	◇	◇
Manage performance review process	◆	◇	◇	◇	◇	◇
Provide resources	◆	◇	◇	◇	◇	◇
Manage human resources		◆	◆	◇	◇	◆
Manage facilities		◇	◇	◇	◆	◇
Assure working environment		◆	◇	◇	◇	◇
Develop teaching and training plans	◆		◆	◇		◆

(*Table 6.1 Continued*)

(Table 6.1 Continued)

QA duty	Office of quality assurance	Office of administrative affairs	Academic departments	Office of students affairs	Technology office	Training office
Attend to students' needs and demands		◇	◇	♦	◇	♦
Design and develop courses	◇	◇	♦	◇	♦	◇
Purchase goods and equipments		♦	♦	◇	♦	♦
Carry out teaching and training		◇	♦	♦	♦	♦
Develop assessment and improvement plan		♦	◇	◇	◇	◇
Monitor and assess quality performance		♦	◇	◇	◇	◇
Control under- or unqualified performance		♦	◇	◇	◇	◇
Analyze monitoring/ assessment data		♦	◇	◇	◇	◇
Make improvements		♦	◇	◇	◇	◇

Source: Liu (2005).
Note: ♦ Main responsibility; ◇ Auxiliary responsibility

ACHIEVEMENTS

As a result of the comprehensive efforts discussed above, SDLME first obtained ISO 9001:2000 quality management system certification in March 2003. In April 2006, SDLME passed the ISO 9001:2000 reevaluation audit and earned the second round quality management system certification. In April 2009, SDLME passed the third audit and obtained ISO 9001:2000 quality management system certification for the third time. In addition, in March 2010, SDLME obtained ISO 9001:2008 quality management system certification.

These achievements indicate that SDLME has successfully operated the ISO 9001:2000 quality management system by:

- Incorporating its standard concepts and methods into the daily work of all staff;
- Developing an institution level QA framework and standards;
- Standardizing quality management procedures;
- Applying the eight quality management principles (customer-centered, leading role, total employee involvement, process approach, systems approach to management, continual improvement, factual approach to decision-making, mutually beneficial supplier relationships) in its QA activities; and
- Promoting the "Plan–Do–Check–Act" (PDCA) cycle at all levels (Liu, Gao, and Sun 2003).

CHALLENGES AND LESSONS LEARNED

ISO 9000 as a Tool to Produce Organizational Changes

The SDLME case has taught us that the ISO 9000 quality management system can be used as a tool to bring about changes in quality concepts, mechanisms, and processes, not just as a way to obtain certification. After observing the use of ISO 9000 at many open universities to improve the quality of DE programs and services, SDLME has introduced the ISO's QA system and found that ISO 9000 has helped SDLME to:

- Establish an internal QA system;
- Pay close attention to distance learners by emphasizing the analysis of learners' demands before making any QA judgments. Staff gradually develops an awareness of the value of demand-oriented decisions, which is the basis of QA;
- Consider the DE market, institutional capacity, and resources before introducing new programs, courses, or services. When China's modern DE project was first initiated, the pilot colleges hastily offered too many programs without fully considering prospective students' needs and securing physical and human resources to meet those needs, which led to poor quality online education (Deng and Feng 2004); and
- Realize the importance of self-improvement as more than just conforming to external quality requirements. This realization did not come easily. At the initial stage, SDLME staff felt uneasiness and stress from being required to follow the ISO 9000 quality management system. During the process of applying the system in their daily work and routine, staff became aware that their QA activities lead to internal quality improvement as well as public accountability, which has contributed to the creation of a quality culture within SDLME.

ISO 9000 as a Tool to Promote Process Analysis

The SDLME experience has taught us that ISO 9000 should be used as a tool to understand our work better and develop actions based on that enhanced understanding. ISO 9000 begins with identifying all detailed procedures of a certain work process, and recording the outcome in a written document. That is, all detailed work procedures analyzed during the identification process are saved as a document file to be used for the next action cycle. All actions are carried out based on knowledge of the required work procedures.

ISO 9000 is flexible, permitting SDLME to develop its own set of self-improvement measures within a framework of continual quality improvement. It emphasizes the process of evaluation and analysis of inputs, using a variety of tools including:

- Satisfaction survey results, statistical analyses, and market investigations;
- Exploratory analyses of corrective actions and preventive measures;
- Internal audits; and
- Evaluations.

This process allows an organization to transform itself, undergo organizational improvement, and develop action strategies based on new work procedures informed by changes in the surrounding environment, market demands, and new levels of understanding.

CONCLUSION

From the macroscopic perspective, China needs to establish a national QA policy framework for DE in order to continuously ensure the quality of DE institutions and their programs. Both accreditation and academic audit models should be included in this QA framework so that all DE institutions, even the well-established, undergo periodic accreditation, and academic audit. Within the national QA policy framework, detailed QA standards that promote continuous quality enhancement should be developed to support distance teaching and learning processes.

From the micro perspective, DE institutions can apply the ISO quality management system to help establish an internal QA mechanism. Using this system, a DE institution can standardize its work flow, perform quality management activities and improve organizational efficiency and effectiveness.

However without a national QA system, ISO cannot ensure consistency of quality standards between different DE institutions.

REFERENCES

Deng, X., and Feng, L. 2004. "The enlightenment of ISO 9000 Accreditation in Distance Education," *China Distance Education*, 5: 6–7.

Liu, Y. 2005 Unpublished Master Degree dissertation. *Action Research on ISO 9000 Quality Management System in Distance Education*. Beijing: Beijing Normal University.

Liu, Y., Gao, S., and Sun, B. 2003. "ISO 9000 Quality Management System in the Practice of Distance Education," *China Distance Education,15*: 43–46.

Peking University School of Distance Learning for Medical Education. 2011. Available online at http://www.pkubytime.com.cn/english/ english.htm and accessed on September 4, 2012.

Mongolian e-Knowledge 7

Sanjaa Baigaltugs

INTRODUCTION

While some rudimentary forms of distance learning existed earlier, it was not until after 1992 that national institutions and nongovernment organizations (NGOs) in Mongolia began to offer distance education (DE) programs. Since then, DE, especially e-learning, has found wider usage in the field of higher, formal, and vocational education in Mongolia as available technologies, including the internet, have greatly facilitated teaching and learning methods. During this process, the Mongolian government has made special efforts to collaborate with governmental and nongovernmental organizations from other countries and regions such as the UK government, the European Union, UNICEF, the Soros Foundation, the Asian Development Bank (ADB), and many others to promote the use of technology in education.

The main planning and implementation agency for quality assurance (QA) and the accreditation of higher education in Mongolia is the Mongolian National Council for Education Accreditation (MNCEA). Established in 1998, the MNCEA conducts accreditation based on evaluation at both institutional and program levels. It functions by developing procedures to assure quality in higher education institutions (HEIs) and providing them with technical assistance and consulting services to improve their overall performance. Most of the public and private HEIs in Mongolia have their programs accredited by the MNCEA. The MNCEA has not developed separate QA/accreditation standards for DE/e-learning, even though the accreditation of such programs comes under its jurisdiction. It is noted that in Mongolia the amount of public funding received by the institution is not dependent on its accreditation status. Hence, accreditation is not generally given serious consideration by many Mongolian HEIs.

DEVELOPMENT OF DISTANCE EDUCATION AND NGO INVOLVEMENT

Since 1998, more than 10 DE projects have been implemented by a number of national HEIs and NGOs. Most of these projects are funded by international agencies such as the World Bank, ADB, the International Development Research Centre (IDRC) of Canada, and Japan-Funds-in-Trust. The following projects are related to DE or use ICT in education.

1. *Knowledge Network*, funded by the IDRC and supported by the Ministry of Education, Culture and Science (MOECS), was initiated in 1998 by an NGO, the Internet, and Information Center, with the aim to provide news and information for teachers and students through its website. Under this project, two schools (one in Ulaanbaatar and another in a rural area) were provided with internet connections.

2. *ThinkQuest* was introduced in Mongolia between 1998 and 2002 with funding from the Mongolian Foundation for Open Society (MFOS). ThinkQuest, a global learning platform, fosters collaborative learning and cooperation among students and teachers from 80 nations around the world and is sponsored by the Oracle Help Us Help Foundation. As a ThinkQuest national partner, the Mongolia Development Gateway organized local competitions, supported the participation of selected Mongolian teams in the international competition and provided professional consulting on content development.

3. *Education Sector Development Program*, funded by ADB, was implemented by MOECS in 1998. It contributed to furnishing over 90 secondary schools in rural and urban areas in Mongolia with computers, offering training courses for ICT education teachers, and providing technical support.

4. *Internet-based Distance Education Project*[1], initiated and funded by the IDRC between 2001 and 2002, introduced

---

[1] http://www.elearning.mn

internet-based DE methodologies to selected Mongolian learning communities. Outcomes of the project included several Web-based courses on specific subjects such as English language, ICT and computer skills, gender issues, and legal rights. A Mongolian consulting company, InfoCon Co., Ltd., was responsible for managing and coordinating all activities of the entire project implementation, and developing training materials in ICT. In addition, it recommended national strategies to develop technology-based DE, based on its project experience.

5. *Mongol Education*[2] is the first educational portal site providing a variety of educational and training resources for teachers and students in Mongolia. Funded by MFOS and managed by InfoCon Co., Ltd., it provides educational information to the Mongolian education community and contributes to increasing public awareness of education reform issues.

6. *Video conferencing centers* across 12 rural provinces have been established and facilitated jointly by the ICT Training Center and MOECS within the framework of the "Capacity Building for Civil Servants," a project funded by ADB. It is anticipated that these video conferencing facilities will be utilized for the training of postgraduate teachers through distance mode.

7. *E-learning Center* was established at the Computer Science and Management School (CSMS) of the Mongolian University of Science and Technology (MUST) in 2003, with support from the Centre for International Cooperation for Computerization of Japan. Its mission has been to develop online content for DE. So far, more than three hundred content modules have been developed in areas of engineering study and used by learners from MUST.

Whilst these developments concern various DE projects, in 2000 the Mongolian government adopted *ICT Vision 2010* as a blueprint for ICT development in the country (Government of Mongolia 2000). The major activities undertaken in *ICT Vision 2010* included:

---

[2] http://www.mongoleducation.mn

- The development of a structure to provide ICT education to all citizens;
- The establishment of high-tech centers in Ulaanbaatar and major cities of other socio-economic development areas to offer ICT education;
- The creation of infrastructure for education;
- The development of a detailed human resources development plan to meet ICT; competency requirements for users, trainers, and specialists;
- The development of electronic library systems;
- The promotion of lifelong learning through open and distance learning; and
- The introduction of electronic services for leisure and entertainment activities.

Further, in 2002 the Mongolian government approved a *National Program on Distance Education* with the main goal of providing its people with an opportunity to engage in lifelong learning for the improvement of their living standards, and to build a national DE system (Uyanga 2005). The objectives of this Program included:

- The establishment of a national DE strategy, coordination, and management;
- The creation of a mechanism for DE services and activities;
- The development of human resources capacity to train DE specialists;
- The creation of a quality distance learning environment; and
- The selection of the most appropriate methods of distance learning to deliver content.

In 2005, the *ICT Vision 2010* was further elaborated in the *E-Mongolia National Program* to establish a knowledge-based society in Mongolia by enhancing ICT applications in all sectors of society (Government of Mongolia 2005). In education, the *E-Education Project* was initiated:

- To achieve an average international ICT literacy level of 80 percent of all capable people by 2012;

- To ensure that 70 percent of all "soums" (a primary administrative unit like a county) and all provincial capital cities will have access to the DE system by 2012; and
- To create a model "e-school" and ensure that 50 percent of schools will have e-school capability by 2012.

Since 2006, many more ICT development projects, including online and mobile content development and the expansion of fiber optic cables, have been implemented. However, as Bates (2011) notes, DE in Mongolia remains new or underdeveloped and national policies to support the development of quality DE and effective ICT integration in education are still lacking.

In recent years some NGOs and HEIs have been collaborating with foreign partners in developing and delivering quality e-learning programs in Mongolia. The following section introduces one such case and highlights QA efforts with respect to a strategic partnership with an established foreign e-learning provider.

MONGOLIAN E-KNOWLEDGE AND ITS QUALITY ASSURANCE

Overview

The NGO *Mongolian e-Knowledge* (MeK) was established in 2007 to promote e-learning in Mongolia through the use of ICT. Its main objective is to contribute to the country's development by providing opportunities for individuals and organizations to share and manage knowledge using electronic means. MeK started its involvement in e-learning by partnering with a nonprofit organization in Germany, *Inwent*[3], on October 23, 2007. Under this partnership, MeK staff received formal training from Inwent to serve as online tutors.

Within Mongolia's *ICT Vision 2010* framework, MeK was tasked with developing and delivering e-learning courses throughout the country. MeK has developed certificate courses such as

---

[3] http://www.inwent.org/index.php.en

"Instructional Design," "Tutoring in e-Learning Communities," and "e-Learning Content Development" in the Mongolian language and offered these courses via its own learning portal[4], which is based on the popular open source learning management system, Moodle. Each course is delivered online over four to six weeks. To date, more than one hundred and fifty students have successfully completed these certificate courses. In addition to these certificate courses, several other online courses have been developed. Examples of these courses include e-Learning Strategy, e-Learning Project Management, Support of Virtual Learning Communities, Content Development, and e-Learning Technology. MeK also provides technical support for the course delivery activities.

In early 2011, MeK convened a round table discussion involving participants from ministries, agencies, and other NGOs in Mongolia to develop the first nationwide e-Learning strategy. The discussion group identified several aspects involved in assuring and managing the quality of e-learning. In this initiative, Inwent played a significant role in the introduction of a quality improvement scheme called the "Open Certification for E-Learning in Capacity Building Check" or "Open ECB Check"[5] in MeK. The Open ECB Check, co-developed by the European Foundation for Quality in E-Learning[6] and Inwent, supports e-learning organizations' capacity-building to measure how successful their e-learning programs are and allows for continuous improvement though peer collaboration and benchmarking.

MeK's QA Mechanism

MeK's QA mechanism follows Inwent's quality management system, which was developed by the European Foundation for Quality Management[7] (EFQM). Under the Inwent-MeK agreement, MeK courses adopt instructional design strategies and course content used in Inwent courses, although MeK courses are delivered in the Mongolian language. New courses are developed

---

[4] http://www.mongolcampus.org
[5] http://www.ecb-check.org/
[6] http://www.qualityfoundation.org/
[7] http://www.efqm.org/en/

to meet the demands of the Mongolian market, following Inwent's quality standards. Developers of these new MeK courses include staff from private and public universities and colleges in Mongolia.

QA in Course Design and Delivery

MeK's QA system adopts a holistic systems approach when designing its courses. Academic quality standards in course design are established, and approval and review procedures are specified. Prior to the beginning of each course, a face-to-face orientation meeting is organized to help the MeK team develop the course syllabus and become familiar with its e-campus functions, such as how to enter a training site and participate in discussion forums and chatting. During the course, regular and systematic monitoring of learner support services is conducted to facilitate continuous improvement. After the course is over, a process evaluation report is compiled and used to make adjustments and improvements to the course.

QA in Learner Support

The MeK team constantly strives to improve the quality of learner support services that currently include online consultations, asynchronous communication via email feedback, and synchronous support via telephone and chatting. Free ICT training courses are also provided to learners who demonstrate a lack of ICT skills in using Moodle, before a class begins. In addition, counseling is provided by dedicated guidance teachers on campus to those MeK learners who need to be motivated and helped to keep up with their course schedule.

QA in Evaluation and Assessment

An important component of MeK's QA process is the development of survey questionnaires to evaluate learner satisfaction with regard to how the courses are delivered, how the learning materials are received and utilized, how much time is taken to complete each course, and other quality elements considered important by the learners. Such surveys are administered at the end of each course using internet-based online survey tools. Survey results provide useful feedback on the courses as well as the students' perceptions of the quality of the course content.

Quality as Process

As part of the aforementioned Inwent–MeK agreement, all courses from MeK must comply with the "Quality as Process" (QaP) management system. The QaP system helps the entire organization to be involved in the development process of both individual and group learning environments, and provides feedback about the effectiveness of all activities during the process. MeK also adopts Inwent's planning, monitoring, and evaluation system (PriME) for assuring the quality of its programs. In addition, MeK independently strives to manage its QA process beyond use of the Inwent QA tools by employing self-assessment, external observation, assessment, and verification procedures as tools to assess its productivity. All of these measures are designed to ensure that both positive and negative experiences are considered when strategies or quality policies are updated.

ISSUES AND CHALLENGES

In recent years, the number of students engaged in distance learning has increased as more HEIs in Mongolia offer DE programs. Feedback from the HEIs suggests that Mongolia needs a regulatory QA framework for DE/e-learning provision. Currently, no QA regulatory framework exists, and its absence may lead to poor quality DE/e-learning programs. Policies and guidelines for QA of DE/e-learning systems are needed for the further development of this mode of education delivery in Mongolia.

MeK has encountered several issues and challenges related to its quality management. These include:

- *Lack of internal capacity*: Some staff members are unable to deliver good quality courses due to their lack of understanding of the quality aspects of e-learning. To address this issue, MeK has provided staff training prior to the launch of the courses. However, there is still need for continuous training opportunities, written manuals, and good communication between staff and the organization.

* *Increased competition*: With the increasing number of institutions adopting Moodle to deliver e-learning in Mongolia, MeK faces competition from other organizations. Their courses use the same open source content but are delivered over longer durations, which are more attractive to students. MeK should develop a clear, long-term plan, which includes clear standards for quality courseware development and support.
* *Students' low e-learning readiness*: Many students are unfamiliar with e-learning approaches and have not experienced this form of learning before. Some expect lectures to be delivered, as in the case of face-to-face delivery. This mismatch of expectations of students and providers often results in student dissatisfaction. To meet this particular challenge, MeK has begun to deliver most courses in a blended mode of face-to-face and online education.

LESSONS LEARNED

This chapter has argued that the quality of e-learning can be assured through the management of processes in course planning, design, development, delivery, and evaluation. It has also stressed the importance of monitoring and evaluation to determine compliance with stated requirements, and to achieve the prescribed outcomes and beyond. Moreover, there is an identified need for a dedicated regulatory authority for QA in DE/e-learning in Mongolia to underpin the further development of DE/e-learning in the country.

The MeK experience has taught us that the purpose of QA is to ensure the consistency of services and products, reduce variability and unpredictability, and thus improve reliability in delivery and quality. It has also shown that QA should aim to make processes and procedures clear to the people using them; that is, to promote transparency and reduce uncertainty, and to avoid errors as a consequence. In the context of e-learning, QA should reflect the unique features of e-learning, as e-learning often adopts flexible ICT tools to promote students' active participation. In order to establish such QA systems, stakeholders, organizations, and government agencies must communicate with each other to develop a better understanding of what is necessary to improve the quality of DE/e-learning. As indicated by Daniel (1999), communication, coordination, and careful attention to details are needed for QA in DE/e-learning.

The MeK experience has also revealed that when there is no national QA system for DE/e-learning, a strategic partnership with a successful international non-profit organization with worldwide operations dedicated to human resource development helps DE/e-learning providers build internal capacity. This has been achieved by introducing appropriate QA policies and procedures and developing a quality culture, all in a relatively short period of time. Among more than five thousand NGOs registered with the Ministry of Justice and Home Affairs in 2005, one hundred and eighty three cited education as their main activity. But MeK is the only NGO which adopts QA standards in developing and delivering e-learning. MeK's main advantage lies in its collaboration with an international institution, Inwent of Germany (now GIZ, as of 2011), that offers practice-oriented e-learning with SQS (Software Quality System) certification. Other NGOs in Mongolia who are planning to start DE/e-learning should consider collaboration with experienced organizations from other countries to build an internal QA system and member capacity.

The MeK case has evidenced the gradual development of a local need-based QA system as experiences in QA activities are accumulated. Previously, Inwent courses were simply translated into Mongolian, and the MeK team did not pay too much attention to the QA elements, as the course content was designed by Inwent. But during course implementation, the MeK team realized the necessity and importance of modifying the course content to make it more user-friendly to meet the needs of the target learning groups in Mongolia. MeK's QA system now integrates several measures that reflect Mongolian users' needs within the QA framework offered by its external partner Inwent. The time is ripe for MeK to invest resources and take the necessary action to develop its own QA system. It is envisaged that MeK may contribute to the establishment of a national QA framework for DE/e-learning and the creation of QA criteria and standards in the near future. MeK practice has taught us that capacity-building efforts should be made to improve the quality of DE/e-learning. DE/e-learning is still new in Mongolia due to lack of ICT knowledge and skills, limited high-speed broadband infrastructure, lack of funds for e-learning development, insufficient e-learning practice, and supporting national policies. For these reasons, DE/e-learning will need more time and development before it is ready to play a major role in the provision of higher education to the Mongolian people. So far, universities such as National University of Mongolia, Mongolian University of Science and Technology and NGOs such as MeK have taken steps to establish and develop DE/e-learning programs. Even though there is no national QA system that helps these DE/e-learning providers achieve a high level of quality teaching and learning and ensure quality services and products, some Mongolian DE/e-learning institutions have tried to learn from the QA benchmarks and best practices from other providers in developed countries. Government agencies, NGOs, and state universities involved in DE/e-learning have sent their representatives to other countries to participate in workshops and seminars related to DE/e-learning for capacity-building.

CONCLUSION

In conclusion, the following QA mechanism is proposed to ensure the development of a quality DE/e-learning component of the Mongolian education system.

Policy and Regulatory Framework

- The Government should develop a central policy to promote the wider utilization of quality DE/e-learning and establish a strong accreditation and QA system for DE/e-learning programs offered by national and international providers.
- The development of national QA protocols, criteria, and

standards for DE/e-learning programs must be in line with international best practices and guideline manuals, and reflect DE/e-learning experiences of Mongolian HEIs and NGOs.

- Funding must be made available to support planning and monitoring compliance.
- The capacity of Government QA regulatory agencies dealing with accreditation and monitoring of DE/e-learning practices should be improved to provide effective and efficient QA services.

Staff Development

- Capacity-building and training opportunities must be provided to staff of universities, NGOs, and other organizations offering DE/e-learning to improve the knowledge and skills related to QA activities.
- Seminars and workshops on how to evaluate the quality of DE/e-learning should be organized for assessors and accreditors.
- Continuous support and retraining for staff members should be institutionalized for continuous quality improvement in course development and implementation.

Technology Enhancement

- The National Center for Non-formal and DE in Mongolia must play a major role in introducing and maintaining new technologies for the quality improvement of DE/e-learning under the national plan, such as E-Mongolia.

The Mongolian National Council for Education Accreditation (MNCEA) will be a main policy initiator for a Mongolian DE/e-learning QA system, and evaluator of the quality of DE/e-learning programs offered by various institutions in the private and public sectors. Thus, to enhance the quality of DE/e-learning that is so keenly promoted by the Mongolian Government, the MNCEA should develop QA guidelines and standards in DE/e-learning in order to improve public confidence and acceptance.

The MNCEA should also develop a system to monitor the performance standards so that continuous quality enhancement can take place. Moreover, institutions offering DE/e-learning should be encouraged to join regional bodies such as the Asian Association of Open Universities (AAOU) to benefit from networking with other experienced practitioners as well as adopt QA criteria and standards frameworks developed by AAOU as well as other international/regional organizations.

REFERENCES

Bates, A. 2011, January 14. *Distance Education in Mongolia*. Available online at http://www.tonybates.ca/2011/01/14/distance-education-in-mongolia/and accessed on September 4, 2012.

Daniel, J. 1999. "Open Learning and/or Distance Education: which one for what Purpose?," in K. Harry (ed.), *Higher Education through Open and Distance Learning*, pp. 292-298, London: Routledge and Commonwealth of Learning.

Government of Mongolia. 2000. *Concept of ICT development of Mongolia by Year 2010*.

Government of Mongolia. 2005. *E-Mongolia National Program*.

Uyanga, S. 2005. "The usage of ICT for Secondary Education in Mongolia," *International Journal of Education and Development using Information and Communication Technology (IJEDICT)*, 1(4):101–118.

S. Korea's AutoEver 8

Hae-Deok Song and Cheolil Lim

INTRODUCTION

South Korea ("Korea" hereafter) has witnessed rapid growth in the area of corporate e-learning in recent years. Overall sales in the Korean e-learning industry rose 11.8 percent between 2008 and 2009, reaching US$27 billion, and the number of e-learning providers also increased by 19.5 percent during the same period, totaling one thousand three hundred and sixty eight (Ministry of Commerce, Industry and Energy, Korea Institute for Electronic Commerce and Korea Association of Convergence Education 2010). With the enactment of the e-Learning Industry Development Act in 2004, the Korean e-learning industry solidified its foothold as a leading knowledge service industry, demonstrating 10 percent annual sales growth and 39.6 percent annual growth in the number of providers (Lim 2007). Despite this quantitative growth, however, there has been increasing criticism of the quality of e-learning programs, with the Korea Consumer Agency receiving nearly ten thousand consumer complaints related to e-learning between 2007 and late August 2010. This implies that effective quality assurance (QA) is needed and crucial to the continued advancement of the corporate e-learning market.

Since 2000, Korean companies that adopted e-learning programs have been financially supported through the Employment Insurance Reimbursement Policy (Lee 2005). A prerequisite for eligibility for this support is to demonstrate a given level of quality in an evaluation conducted by the e-Learning Center of the Korea Research Institute for Vocational Education & Training (KRIVET). The Center evaluates whether the e-learning courses/programs meet its QA standards and contribute to the development or improvement of employees' competencies (Lee, Byun, Kwon and Hwan 2006). For this, the Center examines the QA capabilities of e-learning providers in terms of the relevance of their courses/ programs to corporate employees, and the suitability of their courses/programs to meet market needs (Human Resources Development Service of Korea, 2009; Lee, Byun, Kwon and Kwak 2009). The Center's QA areas include course development and delivery, training methods, assessment and feedback, technology and infrastructure considerations, learning management and support, and tutors and other support personnel. The Center's evaluations are strict and thorough, and its QA standards are high. In 2009, for instance, 15 of a hundred and seventy three providers received ratings of A (excellent), 49 were rated as B (average), and 57 providers received C (poor) ratings (Ministry of Labor and Center for the Evaluation of Skill Development Policy 2010). Hyundai AutoEver (AE), a Korean e-learning corporation, displayed remarkable levels of achievement through this evaluation. The cumulated percentage of AE programs with ratings of B or above in the evaluation was 83.8 percent in 2009. This is extraordinary, considering the fact that only 37 percent of all Korean e-learning companies received ratings of B or above in that year.

Hyundai AutoEver (AE) is an IT service company that is responsible for developing and providing programs on office practices, information and communication technologies (ICT), and foreign languages within the Hyundai Motor Group. As a comprehensive IT service provider, AE offers engineering services, system integration services, infrastructure services, consulting services, and e-learning programs exclusively for the Hyundai Kia Automotive Group. The E-learning Unit at AE provided about 10 percent (US$42 million) of its total sales of US$498 million in 2009.

AE was recognized as the Best E-learning Company of the Year for 2008–2009 by the Ministry of Labor and has also received ISO certification, which sets a high benchmark for Korean corporate e-learning providers (Lim 2011). AE's quality achievement, as shown in these external evaluations, owes much to its careful use of a range of QA tools at every stage of its e-learning development and implementation. The aim of this chapter is to analyze these QA tools within the overall QA system at AE, especially its E-learning Unit, and discuss lessons learned from the AE experience.

HYUNDAI AUTOEVER'S QUALITY ASSURANCE SYSTEM AND TOOLS

With a centralized organizational structure, AE uses a variety of QA tools and learner support tools at each stage of its e-learning development and implementation in order to improve the quality of its e-learning programs.

Organizational Structure

AE has established a strong and continuous quality enhancement system. Figure 8.1 shows the structure of the E-learning Unit, which consists of three teams: *a)* a content business team, *(b)* a content service team, and *(c)* a training team. The content business team provides e-learning consulting and marketing services. The content service team operates an entire suite of QA practices, and is divided into five main sections: planning, development, management services, system monitoring, and a help desk. The first three sections (planning, development, and management) take the responsibility for developing and implementing e-learning programs. The system monitoring section monitors service systems, provides answers to system related problems, and improves the e-learning program management system. The help desk section receives feedback and handles complaints about courses to improve their quality. Lastly, the training team provides face-to-face and blended learning sessions.

An external team of consulting professors advises on overall QA issues related to e-learning course development and operation, such as instructional design, tutor management, and system design. QA activities are coordinated by AE's quality innovation team, which establishes QA regulations and creates standardized guidelines for program development methodology.

AE e-Learning Programs and QA Course Evaluation Systems

The AE e-learning programs are primarily divided into two categories: general and automobile-specific. The general programs include courses that are basic to all types of employees. Core areas

Figure 8.1 QA organizational structure for the AE E-learning Unit

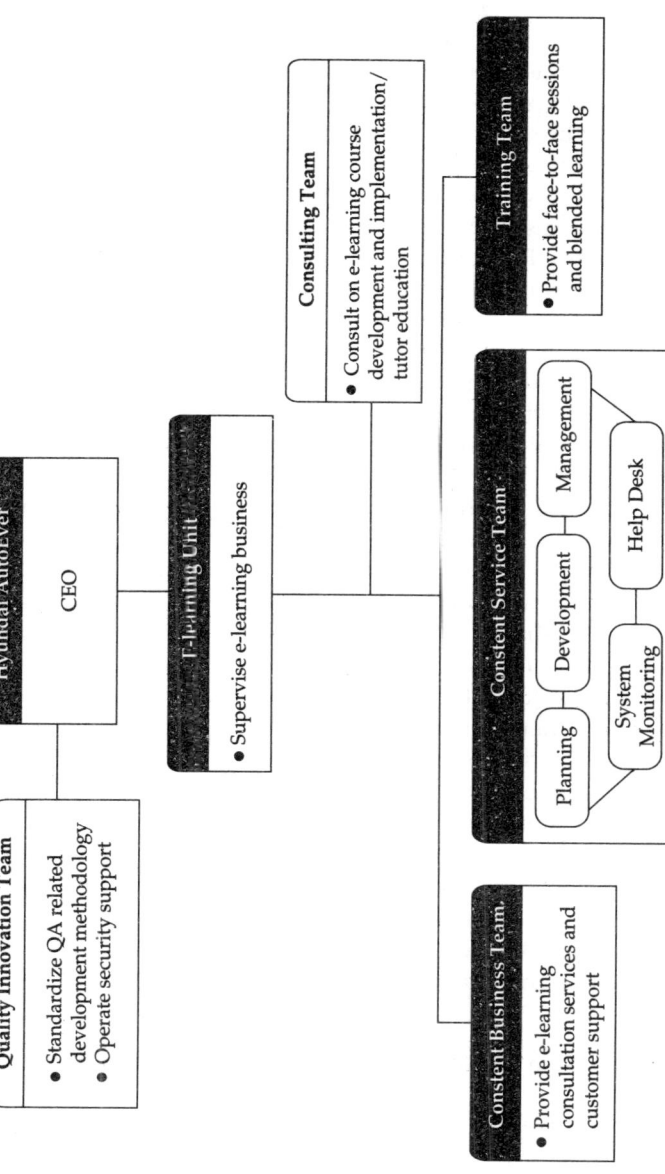

Source: QA organizational structure for the AE E-learning Unit. Adapted from Hyundai AutoEver e-learning quality control manual by Hyundai AutoEver (2009) p. 77. Copyright 2009 by Hyundai AutoEver. Adapted with permission.

include global education, leadership, business foundations, and business management, and each area includes a variety of specific e-learning courses. The global education program, for instance, includes courses such as Global Business Leadership, Global Business Documentation Work, Global Business Negotiation, Business Meetings, Understanding Brazil, Business Opportunities in China, and Introduction to Russia.

The automobile-specific programs cover information and skills related to business, production, sales, maintenance, after service payment, systems, R & D, and steel as they apply to the automotive industry. For instance, the maintenance program includes the following courses: Car Communication Systems, LPG–LPI Engines, Electronic Control Share Systems, Brake Systems and ABS, Car Maintenance Case Studies, Guide to Gasoline Engine Diagnosis, Hybrid Cars, and Car Protection Maintenance.

AE manages the quality of its e-learning programs through its course evaluation system. The company undertakes a course evaluation each year to determine whether existing courses should be modified while retaining their basic forms, upgraded, or removed altogether. Figure 8.2 shows the course evaluation process, which involves collecting data from a variety of sources such as learner satisfaction surveys, learner opinions, regular operational meetings, and learner interviews. Both quantitative and qualitative standards are employed to evaluate the quality of each course. Quantitative standards enable programs to retain courses only when course content has been produced within the last two years and the evaluation results in a rating of B or above, or a satisfaction level of 4.0 or above.

Qualitative standards, on the other hand, consider learners' opinions and face-to-face meeting reports. If a course evaluation results in a score of less than 60 or if a course is judged to be inappropriate for some other reasons, the course is immediately removed. Approved courses, on the other hand, are further upgraded or augmented with more in-depth content.

Individualized Learning Process Management

One important aspect of successful e-learning programs is individualization, the ability to address the differing needs and

Figure 8.2 E-learning course evaluation process

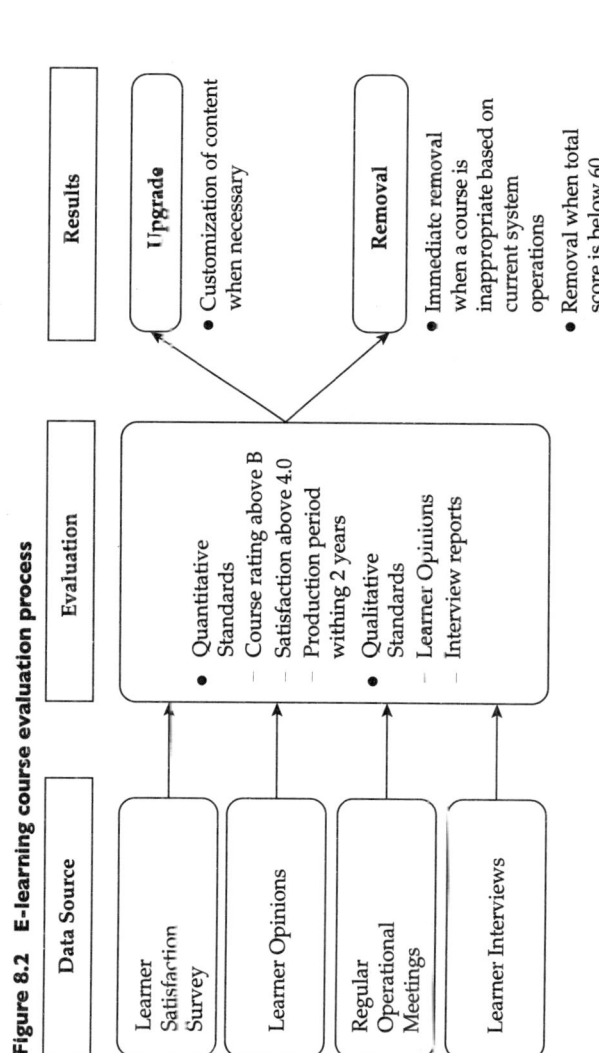

Data Source	Evaluation	Results

Data Source:
- Learner Satisfaction Survey
- Learner Opinions
- Regular Operational Meetings
- Learner Interviews

Evaluation:
- Quantitative Standards
  - Course rating above B
  - Satisfaction above 4.0
  - Production period withing 2 years
- Qualitative Standards
  - Learner Opinions
  - Interview reports

Upgrade
- Customization of content when necessary

Removal
- Immediate removal when a course is inappropriate based on current system operations
- Removal when total score is below 60

Source: E-learning course evaluation process. Adapted from *Hyundai AutoEver e-learning quality control manual* by Hyundai AutoEver (2009) p. 70. Copyright 2009 by Hyundai AutoEver. Adapted with permission.

Figure 8.3 Individualized e-learning cycle in an LMS environment

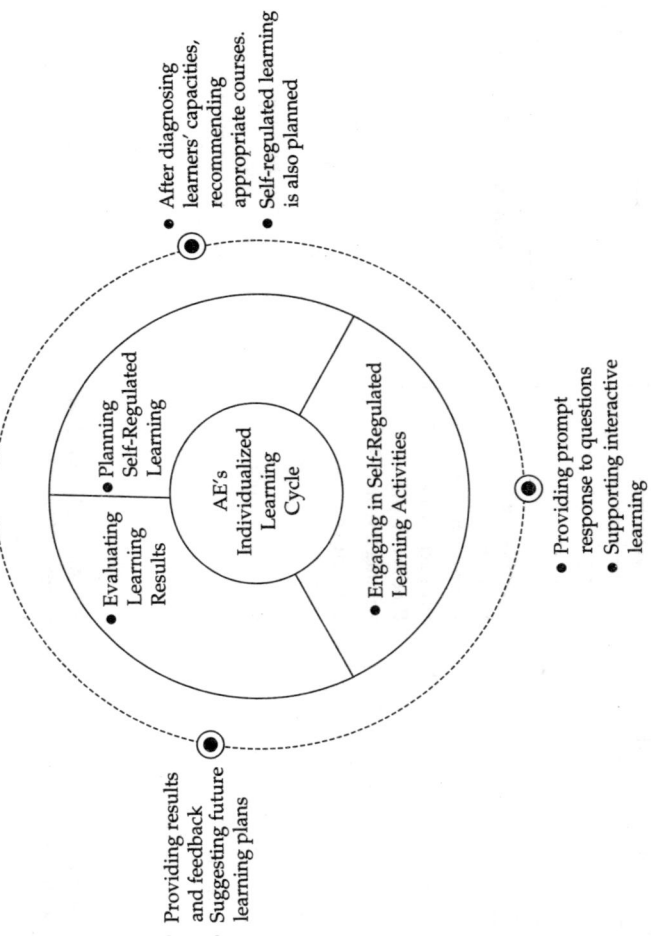

Source: Individualized e-learning cycle in an LMS environment. Adapted from *Hyundai AutoEver e-learning quality control manual* by Hyundai AutoEver (2009) p. 18. Copyright 2009 by Hyundai AutoEver. Adapted with permission.

Note: Adapted from *Hyundai AutoEver e-learning quality control manual* with permission levels of learners. In the past, AE found that e-learning programs that were not individualized required more time for completion on the part of learners, and a greater financial investment on the provider's part. In order to meet this challenge in a cost-efficient and effective way, AE designed an individualized learning cycle (Figure 8.3) that consists of three main phases: planning, self-regulated learning activities, and evaluation. This learning cycle was incorporated into a Learning Management System (LMS), which plays an important role in managing individualized e-learning. With the LMS, program participants first plan their learning based on the results of competency tests, and then undertake a customized learning program that has been designed based on self-regulated learning strategies. The learning patterns of each participant are reported after course completion, and a self-reflective evaluation process is conducted.

QA Tools during the Program Development Stage

AE carefully develops and uses a variety of QA tools during the program development stage to foster effective and efficient e-learning achievement. Table 8.1 presents these QA tools and their uses during each stage of e-learning program development.

In the project planning phase, a variety of tools are used for QA, including:

- Evaluation and selection forms, to increase the validity of adopted training programs and to ensure the quality of course development; and
- Standardized documents for course outlining, proposals, and initial reports, to detect issues that commonly occur in the planning process.

In the analysis phase, an assortment of analysis tools is used for QA, including

- Standard needs assessment forms and requirement questionnaires, to analyze necessary components of e-learning program development that may be missing.

Table 8.1   QA tools and their uses during specific stages of program development

Program development stage (Section in Charge)	QA tool	Use
Planning stage (Planning section)	Evaluation/selection form for outsourcing production	To maintain quality standards for course development in advance
	Course outline	To standardize course planning
	Proposal	To standardize specifications for course development
	Initial report	To standardize outline before the development
Analysis stage (Development section)	Needs assessment form	To standardize development analysis
	Requirements questionnaire	To standardize development research
Design and development stage (Development section)	Basic design form	To standardize design directions
	Detailed guidelines	To standardize detailed design
	Guidelines for the use of the color system	To standardize design
	Prototype (storyboard)	To manage detailed design of pilot production
Evaluation (Management/System section)	Table to identify risk factors	To standardize risk management of project managers
	Discussion of lessons	To share development experiences (successes/failures)
	Education result reports	To standardize learner feedback

Source: Adapted from *Hyundai AutoEver e-learning quality control manual* by Hyundai AutoEver (2009) p. 67. Copyright 2009 by Hyundai AutoEver. Adapted with permission.

In the design and development phase, the following guidelines are used for QA:

- Basic design forms and detailed guidelines, to standardize design directions and processes;
- Guidelines for the utilization of color systems, to establish design standards; and
- Prototypes in the form of storyboards, to examine the validities of pilot products by checking prototypes for functionality and quality.

In the evaluation phase, the following tools are used for QA:

- The risk management experiences of project managers during course development are identified in the form of a table in order to anticipate future risk factors in program development;
- Staff members discuss lessons learned by sharing their stories of success or trial and error; and
- Education evaluation results are reported.

QA Tools at the Program Implementation Stage

QA is also well integrated into the implementation of e-learning programs at AE. Table 8.2 summarizes the QA tools that are used during e-learning program implementation.

Prior to program implementation, the following three QA tools are used:

- Common needs of learners are identified by using a manual of standard responses to frequent questions;
- Learning requirements are identified through quarterly meetings with clients; and
- General information related to the program is offered to instructors through a tutor support manual.

During program implementation, a variety of QA tools are employed:

- Questions posted on a bulletin board are collected and sent to tutors through an automatic mail system;
- Counseling reports that record learners' opinions are provided to teachers in order to help them prepare for program implementation;
- Troubleshooting reports are provided to the staff in order to prevent possible system issues before significant problems develop;
- Help Desk performance evaluation is carried out to examine and modify any quality issues related to program operation; and

Table 8.2   QA tools and their uses during specific stages of program implementation

Stage	QA tools (Person in charge)	Usage
Prior to Program Implementation	Manual of standard responses to telephone calls from learners/frequent questions (Operator)	• To provide manuals for standard responses to telephone calls
	Quarterly meetings with clients (Operator)	• To improve the quality of operating systems through quarterly meetings
	Tutor support manual (Operator)	• To provide guidance on tutoring before the course begins • To provide guidance regarding test grades after course completion
During Program Implementation	Q&A bulletin board (Service team)	• To collect questions and answers on a public Q&A board
	Counseling reports (Operator)	• To record opinions offered by learners during counseling
	Troubleshooting reports (System monitoring team)	• To manage a system before problems occurs
	Help Desk performance evaluation (Service team)	• To evaluate Help Desk workers' attitudes toward work and capabilities
	Operation conference (Help desk)	• To review the course management performance index (satisfaction reports on weekly performance), discuss improvements, and seek solutions to problems
After Program Implementation	Satisfaction survey (Operator)	• To examine satisfaction through responses to questions and find ways to improve the program
	Education results report (Operator)	• To identify education results

Source: Adapted from *Hyundai AutoEver e-learning quality control manual* by Hyundai AutoEver (2009) pp. 33–34. Copyright 2009 by Hyundai AutoEver. Adapted with permission.

- Weekly staff meetings are held to review the course management performance index.

After course completion:

- Satisfaction surveys are provided to clients in order to determine whether they are satisfied with the programs and services they have received; and
- Learning outcome reports are provided through the LMS. Clients can appeal if they think their grades are unfair.

CHALLENGES AND SOLUTIONS

Despite the benefits of using QA tools, AE has experienced many challenges related to quality assurance. Three important challenges can be summarized as follows:

Meeting Unexpected E-learning Demands

One major challenge involved in developing and delivering e-learning programs is that unexpected demands can come from managers, customers, or learners at any moment. AE has been made aware of the need for new courses through these varying demands. This has led to the development of a course evaluation system focused specifically on unexpected demands. As shown in Figure 8.4, the development of a new course evaluation system is initiated when a learner or a partner company expresses complaints about existing programs, or suggests the development of a new program. Error rates and learners' opinions regarding the management of their courses are also frequently monitored. If, for any reason, there is a need for change, either internally or externally, the QA committee makes immediate modifications, such as repairing errors or adding content. As a result, about 20 percent of AE courses are under development. Through this evaluation system, AE has been able to establish a flexible QA system by way of highly flexible course management.

Figure 8.4 E-learning evaluation process for unexpected demands

Data Source	Weekly Review	Modification

Learners' Complaints		Maintain course content if there are only simple errors
Partners' Needs	• Share issues encountered • Review learner satisfaciton • Discuss/possible improvements	• Change or add course contents if the need is significant
Error Rates		

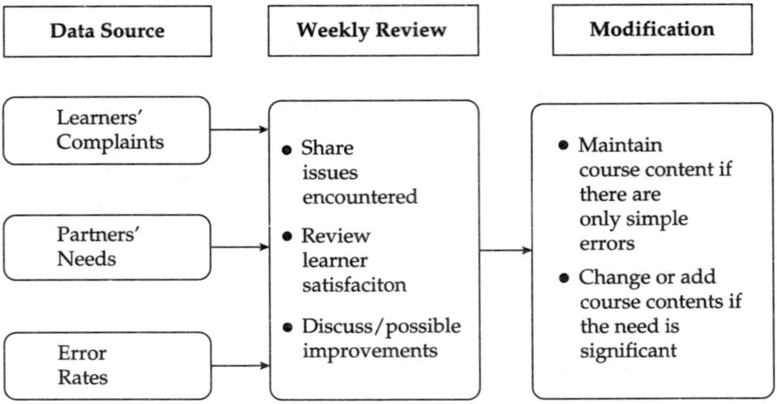

Source: E-learning evaluation process for unexpected demands. Adapted from Hyundai AutoEver e-learning quality control manual by Hyundai AutoEver (2009) p. 70. Copyright 2009 by Hyundai AutoEver. Adapted with permission.

Learner Support Tools to Promote Individualized Learning

To facilitate the individualized learning process, AE has incorporated three support tools. In the beginning, AE used only one LMS to provide e-learning programs. However, complaints from learners increased as the number of courses offered expanded. For instance, learners had difficulty selecting courses as the number of courses grew. They also had difficulty completing courses due to their inability to self-regulate, and they had difficulty with troubleshooting as more multimedia technologies were integrated into the program. In order to meet these challenges, AE integrated three important support tools into the LMS. Table 8.3 presents examples of learner support tools that are included in the LMS: the GLOBE system for competency diagnosis, the SRL system for self-regulated learning, and the MUST system for the detection of learning disorders.

Table 8.3 Learner support tools in LMS

Support tool	Features & Usage	Procedure
GLOBE	• Features: ability to diagnose learners' competencies and recommend customized courses according to the results • Expected effects: improvements in learners' competencies due to customized courses	Learner → Register personal information → Take a competency survey → Confirm diagnosis results → Recommend courses
SRL	• Features: capacity to confirm competency survey results, confirm recommendations based on those results, identify the characteristics of successful learners, and set individual goals and plans for self-regulated learning • Expected effects: shift from unplanned learning activities to planned learning activities	Learner → Confirm competency survey result → Confirm advise → Identify successful learners' characteristics → Set individual goal and plan individual learning
MUST	• Features: ability to diagnose the learners' difficulties in order to prevent possible problems with self-regulated learning • Expected effects: focus on learning	Learner → Select a course → Examine learning environment → Diagnose difficulties in self-regulated learning → Modify difficulties in self-regulated learning

Source: Hyundai AutoEver 2009: 45.
Note: Adapted from Adapted from *Hyundai AutoEver e-learning Quality Control Manual* by Hyundai AutoEver (2009) p. 45. Copyright 2009 by Hyundai AutoEver. Adapted with permission.

Table 8.3 also describes these support tools, which were designed to help students:

- Analyze their core competencies,
- Come up with self-directed learning plans, and
- Deal with common learning disabilities that might interfere with the successful completion of courses.

The commonality among these tools is their enhancement of learner-oriented, self-directed, and individualized learning. The extraordinary percentage of AE e-learning programs with ratings of B or above implies that these tools do support high levels of achievement.

Providing Feedback During Individualized e-Learning

Despite its commitment to individualized learning, AE found that learners were not motivated without being monitored constantly. This especially applied to students who could not manage their learning for themselves in an online environment. These students complained that their failure to complete courses was due to a lack of timely feedback. Thus, a major challenge was determining how to provide feedback that would accelerate each learner's individual learning processes throughout the course. AE carefully designed a feedback mechanism by considering the numbers, channels, and timing of the feedback provided. Feedback is delivered to learners via e-mails and text messages prior to the start of a course, once a week while the course is in session, and at three-day intervals at least twice after completion of the course. For instance, a student taking a four-week e-learning module should receive at least seven messages about his or her progress during the course of the module. Learners with low participation rates are encouraged to participate more in their courses via feedback messages. Direct calls are made in order to remind them of the requirements for course completion. The ultimate goal of this individualized learning progress management is to encourage all learners to complete their courses by providing feedback that reminds them in order to keep up with the work.

LESSONS LEARNED

The AE QA case study provides useful lessons for other e-learning providers who are interested in QA policies, systems, and design. To set up a QA framework in an organization such as an e-learning corporation:

- Government-led QA policies, such as the Korean Employment Insurance; Reimbursement Policy, should be established. This will help to stabilize the e-learning industry in the country in question; and
- A structured organizational setting consisting of an internal operation group, an internal support group such as a team of consulting professors, and an external support group should be created.

To successfully develop and implement an e-learning program, systematic QA cycles and tasks should be established as follows:

- QA should be carried out in discernible cycles that include planning, designing and developing, implementing, and evaluating courses; and
- Clarify the responsibilities of relevant staff members during each stage of e-learning program development and implementation in order to increase the effectiveness of a QA program.

QA tools can play a critical role during QA cycles. More valid QA tools should be employed at each stage of e-learning program development and implementation. A variety of tools should be used for different stages, including:

- Evaluation forms and writing guidelines for the project planning stage;
- A needs assessment form and requirements questionnaire for the analysis stage;
- Design guidelines for the design stage;
- Regulations, guidelines, and prototypes for the development stage; and
- Education reports for the evaluation stage.

To further develop QA systems, course management procedures should be diversified. Given that unanticipated needs are likely to increase with further advances in technology, it is necessary to establish a QA system to proactively embrace changing demands.

CONCLUSIONS AND FUTURE DEVELOPMENT

While the AE system provides one of the best examples of corporate e-learning QA, it still needs further improvement. One important future direction for researchers to take is to develop new QA principles for corporate e-learning from an educational perspective. For this, future QA studies for a corporate e-learning context should focus on:

- Developing more pedagogically valid QA indexes for the entire procedure of e-learning program development. Detailed guidelines must be established to facilitate this as well;
- Exploring QA principles from the learner's perspectives. New theoretical bases such as the constructivist learning environment perspective should be explored;
- Testing specific QA tools empirically to determine their effectiveness;
- Developing alternative QA approaches. Alternative approaches could take various forms, such as a rapid prototyping quality management methodology;
- Examining the utilization of QA tools from the viewpoint of users rather than that of developers;
- Creating QA standards, tools, and principles that are optimal for particular organizational settings; and
- Exploring how QA tools and systems function in different organizational settings.

REFERENCES

Human Resources Development Service of Korea. 2009. *Corporate e-learning Manual*. Seoul: Human Resources Development Service of Korea.

Hyundai AutoEver 2009. *Hyundai AutoEver e-learning Quality Control Manual*. Seoul: Hyundai Autoever.

Lee, H. 2005. "Promoting the Knowledge-Based Economy through e-learning," in J. Kim (ed.), *New Paradigm of Human Resources Development: Government Initiatives for Economic Growth and Social Integration in Korea*, pp. 137–157. Seoul: Korea Research Institute for Vocational Education and Training.

Lee, S., Byun, S., Kwon, S., and Hwan, H. 2006. *Evaluation of Internet Correspondence Training Institutes, Year 2005*. Seoul: Korea Research Institute for Vocational Education and Training.

Lee, S., Byun, S., Kwon, S., and Kwak, D. 2009. *E-learning Policy Development Strategies*. Seoul: Korean Research Institute for Vocational Education and Training.

Lim, C. 2007. "The Current Status and Future Prospects of Corporate e-Learning in Korea," *The International Review of Research in Open and Distance Learning*, 8(1) Available online at http://www.irrodl.org/index.php/irrodl/article/view/376/761 (September 4, 2012).

——— 2011. "Quality assurance for e-learning in the South Korean Corporate Sector," in I.S. Jung and C. Latchem (eds), *Quality Assurance and Accreditation in Distance Education and e-Learning: Models, Policies and Research*. New York and London: Routledge.

Ministry of Commerce, Industry and Energy, Korea Institute for Electronic Commerce, & Korea Association of Convergence Education. 2010. *E-learning White Paper*. Seoul: Korea Association of Conversions Education (KAOCE).

Ministry of Labor and Centre for the Evaluation of Skill Development Policy 2010. *Evaluation of Corporate Vocational Education and Training: Corporate e-learning Briefing Kit*. Seoul: Korean Research Institute for Vocational Education & Training.

PART 3

Focusing on Instructional
Design and Pedagogy

Japan's Kumamoto University Online Graduate School[1]

9

Katsuaki Suzuki

INTRODUCTION

The current quality assurance (QA) framework for higher education in Japan is two-fold: initial accreditation upon the establishment of a higher education institution (HEI) and ongoing academic audits. The latter began in 2004, QA for HEIs having relied solely until then on governmental approval of the institution. The governmental approval system or initial accreditation consists of the Standards for Establishing University (SEU) and the Establishment-Approval System (EAS), while the ongoing audit is known as the Quality Assurance and Accreditation System (QAAS). The Japanese Ministry of Education, Culture, Sports, Science and Technology (MEXT) claims that this "three-fold framework," consisting of SEU, EAS, and QAAS, has both the advantage of the prior regulations assuring proper quality in advance, and the checking after-establishment that continuously assures quality, while respecting the diversity of HEIs (MEXT 2009).

The SEU consist of laws and regulations concerning the basic framework of HEIs, such as qualifications for admission, duration of study, organization, the minimum standards for faculty, facilities and equipment, the norm for educational activities in the university, and regulations for taking courses and requirements for graduation. The process of EAS had been very rigorous, as there was no after-establishment mechanism for assessing quality

---

[1] An earlier version of this chapter (http://www.editlib.org/p/37235) was presented at the *Global Learn Asia Pacific Conference* organized by Association for the Advancement in Computing in Education (AACE), March 29–April 2, 2011 in Melbourne, Australia.

until the School Education Law was amended in 2002 to introduce QAAS.

Having a strict approval system can be seen as an example of a Japanese "convoy system," whereby no university shall be approved until it is confirmed that the new university will fulfill its function as expected, just as the established universities. Under the "convoy system," no organization was supposed to be left behind. Thus, once a university was approved by the government, it became part of the "convoy" to be protected from the competitive environment. However, with the decline in the general population and the consequent lower numbers of university applications in the early 2000s, the government had to give up this "convoy system" and let individual universities maneuver for their own survival. In 2003 the EAS was amended to reduce requirements for initial accreditation and simplify approval standards, and in 2004, QAAS was introduced to periodically assess the quality of HEIs.

QAAS requires all universities in Japan to undergo an academic audit and accreditation process every seven years by certified agencies. As of 2011, there are three agencies approved by MEXT entitled to assess four-year universities and colleges: Japan University Accreditation Association (JUAA)[2], National Institution for Academic Degrees and University Evaluation (NIAD-UE)[3], and Japan Institution for Higher Education Evaluation (JIHEE)[4]. The objective of QAAS is to provide a mechanism through which important QA areas such as organizational management and academic activities of HEIs are periodically assessed and accredited. QAAS requires all HEIs to conduct self-evaluation before submitting their report to one of the three agencies. The accreditation outcome is given on a three-point scale: "satisfactory in meeting the standards," "unsatisfactory in meeting the standards" or "pending," with detailed comments for each evaluation item. The evaluation results are immediately disclosed on the agencies' websites for public viewing. Universities under the "pending" category can seek reevaluation by submitting

---

[2] http://www.juaa.or.jp/en/

[3] http://www.niad.ac.jp/english/

[4] http://www.jihee.or.jp/en/

Table 9.1   Results of QAAS by agency

Year	JUAA			NIAD-UE			JIHEE			TOTAL		
	SF	*USF*	*PDG*	*SF*	*USF*	*PDG*	*SF*	*USF*	*PDG*	*SF*	*USF*	*PDG*
2004	33(23)	0	2	–	–	–	–	–	–	33(23)	0	2
2005	25(10)	0	0	4(0)	0	0	4(0)	0	0	33(10)	0	0
2006	46(8)	0	1	10(0)	0	0	16(0)	0	0	72(8)	0	1
2007	50(13)	0	4	38(0)	0	0	37(1)	0	1	125(14)	0	5
2008	39(17)	0	5	11(0)	0	0	53(5)	0	5	103(22)	0	10
2009	54(20)	0	3	37(0)	0	0	67(19)	0	5	158(39)	0	8
2010	56(27)	1	6	24(0)	1	0	79(26)	1	9	159(53)	2	15

Source: Data obtained from websites of the three agencies: JUAA, NIAD-UE, and JIHEE.

Note: 773 Universities and colleges exist in Japan as of 2010. SF = Satisfactory, USF = unsatisfactory, PDG = pending; Numbers in parentheses indicate numbers of SF with conditional approval.

additional information that indicates improvement within three years. As seen in Table 9.1, the results of QAAS show that most HEIs have met the QA standards.

The above-mentioned "three-fold framework" is applied to distance education (DE) as well. While the procedure for initial approval (EAS) and ongoing audit (QAAS) is the same for both conventional and distance education, the detailed quality standards are different, depending on the delivery mode. For example, for conventional education, direct face-to-face instruction is required as the main mode of delivery, but for DE, it can be substituted with print-based materials with feedback to the submitted reports via mail.

DE has officially been in existence in Japan since 1950. Correspondence programs with print materials were the only alternative until the University of the Air (renamed in 2007 as the Open University of Japan, OUJ) was established in 1985 as a dedicated national distance teaching university, using radio and TV broadcasts as the main means of delivery. OUJ offers around seven hundred courses a year to its approximately eighty thousand students across various study fields. It was assessed by NIAD-UE in 2010 and received a "satisfactory" rating.

Since 2002, governmental deregulation and technological advancement have led to the establishment of a number of online programs and institutions (including for-profit). The Shinshu

University's Graduate School of Science and Technology[5], founded in 2002, was the first to offer a full online option to its master's program. The Tokyo University of Career Development[6] was established in 2004 (but unfortunately had to close its door to new students from 2010), the University of Digital Content[7] in 2005, the Kenichi Ohmae Graduate School of Business[8] in 2005, the Kumamoto University's Graduate School of Instructional Systems in 2006, the Cyber University[9] in 2007, and the Business Breakthrough University[10] in 2010.

The Kumamoto University's Graduate School of Instructional Systems (GSIS) first provided an online program in April 2006, becoming Japan's first graduate school to train e-learning specialists through e-learning (Suzuki 2009). Although the Kumamoto University, founded in 1887, is an on-campus university with about ten thousand students and one thousand academic staff across seven colleges and nine graduate schools, GSIS was created as its first fully online program. It is one of the University President's special projects to investigate the potential of e-learning in higher education, and thus is of an experimental nature. As of April 2011, it has 60 degree-seeking (including 14 in the doctorate program) and 36 nondegree-seeking students who are scattered across Japan. Most students work full-time in a corporate training or higher education sector. The admission quota is controlled by the government at 15 per year for a master's program and three per year for a doctorate program, with no limit in accepting nondegree-seeking students.

GSIS offers four areas of study to train well-rounded e-learning professionals:

- Instructional design (ID);
- Information technology (IT);
- Instructional management (IM); and
- Intellectual property (IP).

---

[5] http://cai1.cs.shinshu-u.ac.jp/xoops/
[6] http://www.lec.ac.jp/english/
[7] http://www.dhw.ac.jp/en/
[8] http://www.ohmae.ac.jp/gmba/
[9] http://www.cyber-u.ac.jp/
[10] http://bbt.ac/

GSIS's master's program, which is equivalent to any other regular on-campus master's program, requires two years of study and a minimum of 30 credit-hours of courses. To complete the master's program, students take 12 required courses and four or more from a list of 16 elective courses. GSIS added a fully online doctoral program in April 2008.

QUALITY ASSURANCE MEASURES AT GSIS

The overall QA process for GSIS is shown in Figure 9.1. Major steps within its QA procedures are explained below.

GSIS and the Establishment Approval System

GSIS experienced the Establishment Approval System (EAS) of MEXT in 2005. As mentioned, EAS is the first QA step for every new higher education program in Japan. Once the application for the establishment of GSIS was submitted to MEXT, an external evaluation team was organized to judge the quality of the application. The application included the reason why the new program was necessary, goals of the program, the curriculum structure and qualifications of teaching staff (professors' academic records), and planned teaching methodologies and facilities to guarantee the instruction would meet the standards required by a regular on-campus equivalent program. Following suggestions from the evaluation team, the application was revised. The subsequent approval enabled the GSIS to start its program in April 2006 under provisional status. As such, GSIS was required to submit end of year reports for any changes in its staff or curriculum for the approval of MEXT, and to receive visits from an external verification team if any concerns were raised (this did not occur). GSIS obtained full status as an online graduate school after two years of provisional status for the master's program and three years for the doctorate program.

Conducting Self- and External Evaluation

The Kumamoto University as a whole conducted self-evaluation and received external evaluation in 2009, as a required practice

Figure 9.1 Overall QA process for GSIS

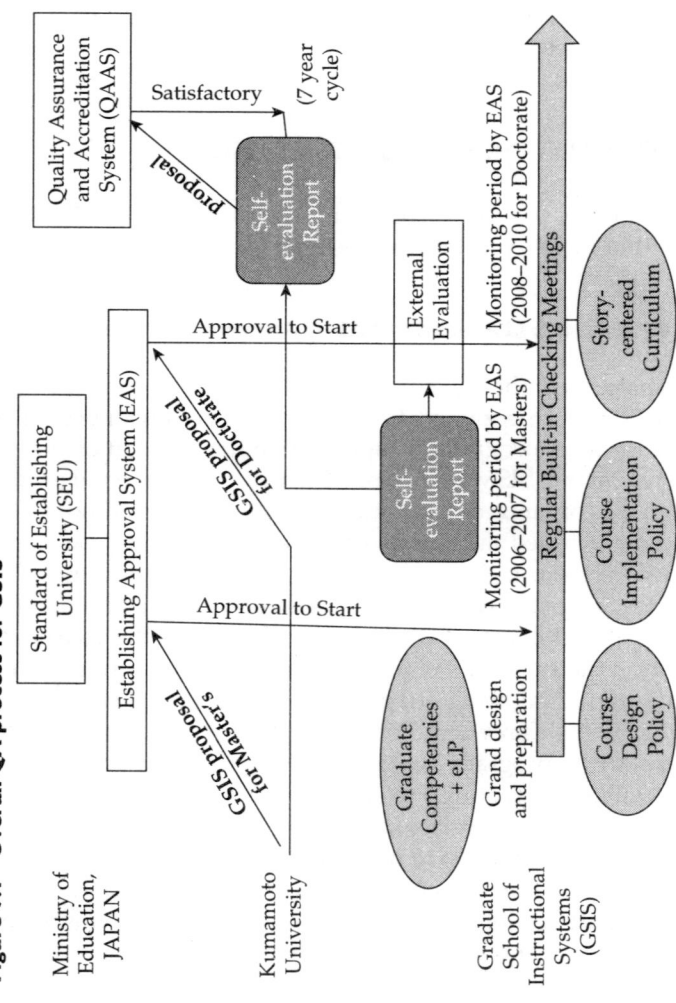

Source: Author.

by all HEIs in Japan since 2004. After two years of self-evaluation, the subsequent report was submitted to an accreditation agency, NIAD-UE, to be reviewed for QAAS. After thorough evaluation of the report and a site visit, NIAD-UE gave the University a pass as an accredited HEI with detailed feedback comments on strengths and areas for future improvements. The report was posted in its entirety in Japanese on both the University's official website[11] and the NIAD-UE's website.

Prior to the University-wide QAAS, GSIS decided to seek an external review in March 2008 in order to:

- Prepare for the University-wide self-evaluation;
- Invite external experts' views and ideas to identify areas for future improvement of its program, since it was approaching the completion of its two-year master's program; and
- Publicize the quality of its program by being evaluated positively by external experts.

Following the QAAS procedure, GSIS formed a self-evaluation committee within the program, compiled a self-evaluation report, and invited an expert panel of external evaluators to examine the report and observe the online courses. The panel evaluated the educational quality of the program very highly, and recommended that more resources should be allocated to faculty research activities, in order for GSIS to produce more research papers on its innovative practice.

Built-in Course Team Meetings as Faculty and Staff Development Opportunities

Continuous faculty and staff development is essential for assuring quality in higher education. In Japan, faculty development at an institutional level became mandatory for all HEIs in 2009, as a function of continuing effort in assuring quality. MEXT does not check the details of faculty development, but mandates the establishment of an organizational mechanism for faculty development when the university goes to QAAS. In reality,

---

[11] http://www.kumamoto-u.ac.jp/daigakujouhou/uneisoshiki/hyouka.html

faculty and staff development is piecemeal, generally consisting of occasional seminars by outside experts, peer observation of classes, offering best teaching awards, or receiving student evaluation of completed courses, which may not lead to substantial improvement in the quality of teaching and learning.

Unlike other Japanese HEIs, GSIS has firmly institutionalized an internal mechanism for checking course quality while the materials are under development (Kitamura et al. 2007). Adopting a Course Team approach, GSIS schedules regular meetings among faculty and staff during preparation of a course. The frequency of the meetings depends on the demands for reviews, ranging from twice a week to once every two weeks. During the meetings, course design and implementation strategies are explored and various ideas for effective online teaching are shared among faculty and staff (examples are shown in Table 9.2). These meetings provide an excellent opportunity for continuous faculty and staff development and have led to the development of two policies:

Table 9.2 Sample decisions made during Course Team meetings

- Decide the top page design which specifies common information across all courses:
  - o   Decide how to use table of contents in each course and have a cohesive appearance;
  - o   Change the unit structure from series to parallel format to allow more learner control;
  - o   Change the method of judging unit pass from report (manual evaluation) to quiz (automatic) format; and
  - o   Increase or reduce task loads of each course to maintain a balance with other courses.
- Increase or reduce task loads of each session to keep good weekly work balance.
- Make fine adjustment of starting and ending points of prerequisite and continuing courses.
- Align assignment deadlines for parallel courses.
- Align student group configurations for simultaneously taken courses:
  - o   Align course contents to facilitate transfer of learning from one course to another across parallel courses; and
  - o   Gradually introduce a variety of group work patterns, from simpler ones at the beginning stage to more complex ones at later stages within a semester.

Source: Extracted from Kitamura et al. (2007).

1. The *GSIS Course Design Policies* to be used as a reference for QA during course design and development; and
2. The *GSIS Course Implementation Policies* to be used as a QA reference when a course is in session.

Implementing Course Design and Implementation Policies

The *GSIS Course Design Policies* were formulated through discussions during our Course Team meetings, reflecting the results from our target audience analysis, governmental regulations and the Kumamoto University's general grading policy (Suzuki 2009). As seen in Table 9.3, interaction is asynchronous, through use of the learning management system's quiz, report submission and bulletin board functions. Instead of having deadlines every week, the policy requires 15 tasks to be clustered into three to five blocks, and, within each block, all tasks are due on the same date (see Policy 4 in Table 9.3). If due dates are set every week in each course, then a student taking four or five courses in a semester would need to handle four or five tasks each week. By introducing a clustered submission policy, the student is allowed to finish several weeks' worth of tasks for one course before moving to the tasks of a different course.

Table 9.3   GSIS course design policies

1. One course consists of fifteen interactive sessions with evidence, e.g., quizzes, mini-reports, answers to practice exercises.
2. Course grades are based not on the score of the final examination, but on multiple reports/products, each of which should be a minimum of 60 percent to obtain a passing grade.
3. Assignments in a course should be directly linked to the competencies specified for the graduates of the program.
4. Due dates for 15 session tasks are to be clustered into 3–5 blocks to enable learner's intensive study and allow flexibility in learning pace.
5. Synchronous whole-class activities should be limited to a maximum of twice a semester per course.
6. Students should be given opportunities to comment each other's reports/products for improvements before final submissions.
7. An introductory video message in all courses or all blocks of a course is provided as a motivator, not as a primary mode of information provision.

Source: Taken from Table 3 (p. 475) of Suzuki (2009), with permission.

While the *Course Design Policies* were set to maintain the quality of the course content, the *GSIS Course Implementation Policies* were developed later to guide the course implementation phase, including such policies as:

- All assignments should be graded within one week after the due date; and
- There will be no "fail" given to any assignment, but "revise and resubmit" be given instead.

These policies provide a ground rule for full- and part-time faculty members, and a continuing effort has been made in monitoring the proper implementation of the policies, as they are a key to assuring the quality of online teaching and learning at GSIS.

Adopting a Competency-based Approach to Promote Outcome-based QA

Believing that the critical goal of assuring quality of course design and implementation is to develop the capacities of our graduates as e-learning specialists, GSIS has designed its master's curriculum backwards, beginning from an analysis of a set of competencies that its graduates should develop as a result of studying the courses. Based on 12 core and 6 optional competencies (Suzuki 2009), each assignment in all courses of the master's program is mapped to one or more of these competencies, so that the attainment of course assignments assures student improvement toward building the competencies necessary for graduation.

To ensure the relevance of the program to the requirements of e-learning specialists, all courses are mapped to the competencies set by the e-Learning Consortium (eLC) Japan toward their certificates of e-Learning Professionals, including The Basics, Contents Creator, Learning Designer, Expert, Manager, Consultant, and SCORM Engineer. GSIS aims to produce graduates who are highly capable of meeting the demands of the real world of e-learning. Thus, aligning our program to the existing professional standards is considered a "must." In close collaboration with eLC, our curriculum is designed in such a way that the graduates can obtain three certificates by taking only required courses. By adding designated optional courses, three more certificates may be awarded. During the collaboration process, eLC also prepared its own professional certification programs.

This mutual relationship between GSIS and eLC has served for the benefit of both organizations. GSIS has been able to create high-quality courses for e-learning professionals, and eLC has been provided with high-level professionals needed for the field. GSIS courses provide partial fulfillment toward certificates of e-Learning Professionals when taken as nondegree-seeking courses, and the graduate program has been accepted as qualification for eLC's professional certificates. GSIS and eLC have agreed to periodically update the requirements of the certificates, as well as making adjustment to the GSIS curriculum when modification occurs in eLC's requirements for certification.

Introducing Story-centered Curriculum to Develop Higher-order Thinking Skills

GSIS introduced the Story-centered Curriculum (SCC) as its main teaching methodology in the master's program to its third cohort of students in April 2008. The development was supported by a Good Practice Grant from MEXT for innovative projects that improve the quality of graduate education. The intention of SCC has been to apply the latest design concepts in online education and to promote higher-order thinking skills of adult learners, with the aim to improve the quality of the GSIS program.

SCC is an extension of Roger Schank's instructional design theory, Goal-based Scenarios (GBS). Whereas GBS is a model for designing simulation courses for learning higher-order skills in a virtual environment (Nemoto and Suzuki 2004; Schank 1996), SCC is used for curriculum-level design by providing an architecture for higher scalability, without losing the "learning by doing" nature of GBS (Schank 2007).

SCC unites multiple courses concurrently offered during a given semester with a common cover story from a real-world situation in which graduates of the GSIS program would be expected to work as e-learning professionals. In such an authentic context, students act as if they are already in the situation, but with assistance and information from faculty when needed. SCC sets a structure for low-cost implementation, as compared to GBS, by:

- Using low-end media such as e-mail and video clips on the Web, rather than high-end branching, digitalized video-based scenarios;
- Incorporating existing resources available online, or in textbook format; and
- Relying on faculty (including tutor and peer) and student interactions, rather than preprogrammed-for-all-possibility sequences of computer–student interactions.

The details of GSIS's SCC can be seen in Suzuki et al. (2009).

Offering Initial Student Orientation for Skill Development and Mindset Building

Believing that the quality of e-learning also depends on learners' e-learning readiness, GSIS offers an online orientation course to its new students to help develop such readiness. Once a student is admitted to the program, he or she is given access to the online orientation course, prior to the start of the first semester. The course aims to help the new students understand important features of GSIS, acquire necessary skills to carry out e-learning successfully, develop the mindset for learner-centered e-learning, and thus be better prepared for the new learning environment. It is believed that a unique set of skills and mindset is required for online students to be successful in their learning, and that students can contribute to the program to achieve a higher quality education. The new students come to share this belief and expectation in the online orientation course at the outset of their online experiences.

The online orientation course consists of a mixture of technical and conceptual sessions:

- Technical sessions teach how to set up a personal computer for network access, how to maneuver the learning management system, and how to use the listserve for communication. During these sessions, it is emphasized that to be visible in the online environment, active participation by the posting of ideas is required; simply reading the postings of others is not enough.
- Conceptual sessions cover goal setting (e.g., which of the e-Learning professional certificates is being sought),

coursework planning (e.g., in which courses will the student register in which semester, considering prerequisite relationships among the courses), self-evaluation of competency levels to identify competencies one has already experienced in his/her job, and decision-making with regard to the involvement in SCC.

By experiencing both technical and conceptual online sessions, new students develop correct expectations for their coming studies, thus being ready to meet new challenges that GSIS`s e-learning program may present during their coursework.

ISSUES FACED AND SOLUTIONS EXPLORED

After approval by MEXT, GSIS had to be quick and efficient in making decisions and building an e-learning environment for new students. The *GSIS Course Design Policies* guided the development of high-quality course content. Efforts were made to provide a high-quality learning environment, and thus avoid becoming a "degree-mill" program from which it is easy to graduate, but with little learning achieved. During these processes, several issues were faced.

First, our students' high expectations toward our program continuously challenged us to apply effective, up-to-date instructional design strategies in developing our e-learning courses. As mentioned, GSIS is unique in that it teaches e-learning design and implementation via e-learning. The students are always keen to see what can be done in e-learning when it is designed by professionals who teach instructional design and technology. The students also expect that our development process should be very well organized and effectively managed, and that our products should be of high-quality. To meet these high expectations, GSIS faculty and staff have made a continuous effort to apply what we teach in our curriculum in designing and delivering our courses.

Nonetheless, we have observed a gap between theory and practice in applying e-learning principles in our fully online classes. For example, various types of group work are suggested as effective ways to promote active learner engagement and learning achievement in theory. But, in reality, a majority of our students have not experienced online group work and thus felt incompetent while undertaking group activities, especially during those dealing with difficult tasks. Our solution has been to introduce group activities with increasing degrees of complexity through the semester (Nemoto et al. 2010b). In addition, we offer technical and psychological support to those who encounter problems at the outset of their group learning experience. The *GSIS Course Implementation Policies* play an important role in bridging the gap between theory and practice by offering flexible and adaptable ways of incorporating e-learning theories and models in our program.

Having our students at a distance with few chances to interact directly, we need to rely more on their perseverance, initiation of learning activities, and independent learning skills. This is a challenge for any institution, but more so for a fully online program. Initial strategic efforts for orienting the new students to the online learning environment, which requires different skills and mindset, should be made in order to acquaint them with online learning (Nemoto et al. 2010a). For example, we should teach our students that reading posts on an online discussion board does not inform others that they have read the posts, and thus they should write something back if they wish others to know they have read the posts. In a face-to-face environment, one's presence is apparent to others, but not in an online environment.

LESSONS LEARNED

Some of the lessons we have learned through the above-mentioned case are:

- To establish a built-in mechanism for faculty and staff development, since they are crucial to QA in education.
- To integrate regular evaluation activities into the everyday routine of designing, developing, and implementing the program. Avoid unexpected, ad-hoc evaluation.
- To offer initial learner orientation and continuous support in order to help online students become familiar with the

e-learning environment. Orientation and support should be provided not only for technical issues, but also for conceptual and philosophical issues of e-learning.

- To adopt evidence-based and well-established instructional design strategies to develop better quality e-learning.
- To gain strong support from senior administration. Ensure that at least one member of senior management supports an innovative program, especially if it is not designed for universal application in the organization.
- To be proactive during the QA process by applying strengthened QA measures, such as inviting external reviewers even when it is not required.

CONCLUSION

Going online can be seen as an opportunity to develop a strong and reliable mechanism for assuring quality, since almost everything happening during coursework can be tracked to establish what has occurred. This is not the case in on-campus face-to-face education. The records kept online may not be perfect, as they cannot store offline learning time or thoughtful thinking process during online discussions. But they provide useful evidence for detecting problems in online education and effective strategies for enhancing the quality of teaching and learning online, compared to those records kept for face-to-face instruction where teaching and learning occurs behind the doors of the classroom.

Our journey will continue as we accept new groups of students and face the reality of an ever-changing globalized society. New students may not be the same as those of previous years, which will present new challenges and issues in the provision of our online education. The changing nature of the globalized e-learning market will also require us to continuously update what and how we teach. The most critical element in assuring the quality of our activities is to have a mechanism to continuously monitor and propose revisions to what we are doing. It is this monitoring and revision mechanism that will enable us to continue as an innovative provider of online graduate education in Japan.

REFERENCES

Kitamura, K., Suzuki, K., Nakano, H., Usagawa, T., Ohmori, F., Iriguchi, N., Kita, T. Ekawa, Y., Takahashi, S., Nemoto, J., Matsuba, R., and Migita, M. 2007. "Quality Assurance Efforts at an Online Graduate School to Train e-learning Professionals: A case of Instructional Systems Programme at Kumamoto University," *Journal of Multimedia Aided Education Research*, 3(2): 25–35. Available online at http://www.code.ouj.ac.jp/wp-content/uploads/No.6-04tokusyuu03.pdf and accessed on September 4, 2012.

MEXT. 2009. *Quality Assurance Framework of Higher Education in Japan.* Higher Education Bureau, Ministry of Education, Culture, Sports, Science and Technology (MEXT). Available online at http://www.mext.go.jp/english/koutou/__icsFiles/afieldfile/2009/10/09/1284979_1.pdf and accessed on September 4, 2012.

Nemoto, J., Kubota, S., Migita, M., Matsuba, R., Kitamura, S., Kita, T., and Suzuki, K. 2010a. "Design of Authentic Learning: A Challenge in e-learning Specialist Graduate Programme," paper included in proceedings of Global Learn Asia Pacific 2010, 1237–1242. Penang, Malaysia: Association for Advancement of Computers in Education.

Nemoto, J., Oyamada, M., Shibata, Y., and Suzuki, K. July, 2010b. "Learning Sketch: A learning reflection activity design," paper included in proceedings of ICoME (International Conference on Media in Education) 2010, 206–213.

Nemoto, J., and Suzuki, K. 2004, August. "GBS Checklist for Training Application," paper included in proceedings of the International Symposium and Conference on Educational Media in Schools, 75–82.

Schank, R. C. 1996 "Goal-Based Scenarios: Case-based Reasoning Meets Learning by Doing," in D. Leake (ed.), *Case-based reasoning: Experience, Lessons & Future Directions*. AAAI Press/The MIT Press.

———— 2007. "The Story-Centered Curriculum," *eLearn Magazine*, Feature Article 47–1, Association for Computing Machinery. Available online at http://www.elearnmag.org/subpage.cfm?section=articles&article=47-1 and accessed on September 4, 2012.

Suzuki, K. 2009. "From Competency List to Curriculum Implementation: A Case Study of Japan's First Online Master's Programme for e-Learning Specialists Training," *International Journal on E-Learning*, 8(4): 469–478.

Suzuki, K., Nemoto, J., Oyamada, M., Miyazaki, M., and Shibata, Y. 2009. "Upgrading an Online Master's Degree Programme based on Story-Centered Curriculum (SCC): A Case Study," paper included in proceedings of ED-MEDIA 2009, 591–598.

Open University of China   10

Li Yawan, Yang Tingting, and Niu Ben

INTRODUCTION

Conventional higher education institutions in China are required to comply with the quality assurance (QA) framework specified by the Ministry of Education (MOE). Distance education (DE) institutions also have to undergo external evaluation under the guidance of the MOE and conduct an internal evaluation, as explained in Chapter 6.

The Open University of China (OUC, originally known as the China Central Radio and TV University), together with the Radio and TV University (RTVU) system, which includes 44 provincial RTVUs, nearly one thousand municipal and prefectural RTVUs and three thousand learning centers, comprises China's DE system. In addition, 68 e-colleges operated by conventional universities have been providing ICT-based DE since the initiation of the Open Education Pilot Project in 1999. OUC and the RTVUs also joined this project to increase access to their DE programs.

In 2007, the MOE conducted a summative evaluation of the project across eight areas: *(a)* philosophy, *(b)* learning centers, *(c)* staff professional development, *(d)* DE facilities, *(e)* teaching resources, *(f)* innovative teaching methods, *(g)* academic management, and *(h)* learning outcomes. As a result, DE is acknowledged by the MOE as an independent form of education in the modern Chinese higher education and lifelong learning systems. Since its involvement in the project, OUC has been offering 92 academic programs in science, engineering, agricultural science, medicine, literature, law, economics, management, and education to more than 2.95 million active students. This number accounts for over two thirds of the total number of DE students nationwide. The remaining one-third are those who are enrolled in the 68 e-colleges at the conventional universities. OUC delivers its DE through three networks: an academic management system network established in 1979, a satellite TV network founded between 1986 and 1987, and a computer network established in 1999. The integration of these three networks has become a prominent feature of DE in OUC.

This chapter describes QA processes undertaken at OUC, analyses some of the problems and proposes possible solutions. It concludes with lessons learned from OUC's experience and future challenges.

OUC'S QUALITY ASSURANCE FRAMEWORK

OUC is responsible for overall planning, coordination, organization, and management of DE in the RTVU system. It has developed a QA framework to monitor the quality of the RTVU system, with 35 QA indicators across five QA areas:

1. teaching resources development and management,
2. teaching process management,
3. learning support services,
4. teaching management, and
5. teaching and learning environment (Du, Yang, Yin, and Zhang 2009).

Teaching Resources Development and Management

Teaching resources development and management refers to the planning and supervision of the development and management of various teaching resources, including multimedia materials. OUC prepares all the documents necessary for the RTVUs to develop and manage the stages necessary to produce high-quality teaching resources.

At the program and course development stage, OUC is responsible for proposing new programs to be used by the RTVUs. First, the concerned faculty members at OUC propose a plan for the development of a new program. Second, this plan is assessed, discussed, and modified by a team of external and internal experts. Third, the modified plan for the new program is presented to the MOE at a national meeting. Once approved by the MOE, the plan is implemented nationwide.

To implement the plan, OUC develops a detailed teaching scheme to provide guidance to the RTVUs. The teaching scheme, reflecting features of the RTVUs' credit system, specifies program introduction, teaching process management strategies, media and methods, course setting, assessment approaches, and evaluation methods. It aims to facilitate automatic course selection and individualized study.

In addition to the teaching scheme, OUC provides directions to the RTVUs for the development of syllabi that specify course requirements such as homework, experiments, practice, and social investigation. OUC's directions are used by all RTVUs to develop standardized or unified syllabi for all compulsory, some elective and practicum courses. These directions follow the course teaching requirements issued by the MOE.

OUC also proposes the requirements for each individual course design and formulates the teaching scheme of all compulsory courses. This teaching scheme emphasizes the integration of multimedia materials and Web-based teaching in designing the compulsory courses.

At the resources development stage, OUC develops teaching resources or materials in various media formats for the RTVUs, following standardized procedures specified in *Rules and Procedures for Teaching Materials Construction* which define content, working approaches, schedules, quality requirements, and evaluation criteria.

OUC manages and regulates the teaching resources development for the unified courses that refer to courses used throughout the RTVUs, and implements a project management system for the development of resources. Each procedure during the resources development is strictly monitored by experts in the relevant discipline and by educational technology experts. The Course Team, consisting of well-qualified experts, is responsible for analyzing the teaching content and learner needs, setting course goals, proposing the design ideas and teaching strategies, developing the teaching materials, as well as providing and updating the Web-based teaching content. In addition, OUC provides advice regarding the use of various teaching resources for the unified courses adopted by the RTVUs.

Moreover, OUC establishes a test bank for the compulsory courses for the final examination for each course. The establishment of this test bank is an important aspect of the standardization of examinations.

Finally, OUC is responsible for setting the criteria for the construction of the teaching resource banks and teaching resources of the RTVUs at provincial, municipal, and prefectural levels. It offers the criteria for sharing resources and the teaching management platform of the unified courses. OUC also develops some courses in collaboration with the provincial RTVUs, which serves as a stimulus and helps to avoid the wastage of resources.

Teaching Process Management

Teaching is the process by which an educator influences the students, physically and psychologically, to achieve an educational goal. In DE, teaching is a process whereby the students learn autonomously under the teacher's supervision in a televised and/or networked learning environment. It includes preparation, delivery, learning activities, practice, and evaluation (Yang 2008).

Preparation

OUC is responsible for developing a teaching design scheme for all compulsory courses for the provincial RTVUs. The teaching design scheme includes guidelines for instruction, multimedia usage, teaching process, and evaluation. In addition, OUC proposes guidelines for unified course guidance, effective teaching and learning strategies, and media selection and integration. Another key QA measure that is offered by OUC is teacher training. All RTVU teachers-in-charge are expected to participate in OUC's teacher training on course content, teaching approaches and practical procedures. Finally, OUC is responsible for developing all entrance testing, conducted for each course so that teaching can be more suitable to different levels. The examination results are analyzed by OUC to formulate solutions to existing problems. Newly admitted students must complete a compulsory orientation course which includes skills training for distance learning.

Course Delivery

Various media are used for course delivery in the RTVUs. OUC is responsible for arranging the transmission of audio-visual courses, management of Web-based courses, and publication of various media teaching materials. For the compulsory courses, where a large number of learners are enrolled, satellite TV and the internet are used to broadcast the courses nationwide. OUC provides the teaching materials for the compulsory courses to the provincial RTVUs, which are responsible for broadcasting these materials locally. The learning centers can also use these materials to perform their teaching activities.

During course delivery, special attention is paid to tutoring for compulsory courses. OUC faculty directly tutor students through live broadcasting classrooms, correspondence, telephone, Vertical Blanking Interval (VBI, part of a television signal that can carry information such as closed-caption text), and internet tools such as e-mail and online discussion forums. In addition, OUC provides guidelines for face-to-face tutorials in the provincial RTVUs, and teaching resources for these face-to-face tutorials.

Learning Activities

Even though OUC's quality assurance system focuses on QA in teaching, it also pays close attention to students' learning processes. OUC provides guidelines and suggestions for autonomous and group learning approaches in the teaching documents. Provincial RTVUs are required to help students form learning groups, engage in cooperative learning, exchange their understanding of course content, and cope with loneliness during the learning process in order to achieve the learning objectives.

Practice Teaching

OUC strives to improve the quality of RTVUs' teacher training courses by offering practice teaching opportunities to the students. Practice teaching is a form of professional training designed to help students apply their theoretical knowledge of teaching, strengthen perceptual knowledge and develop practical skills. OUC specifies requirements for practice teaching, which is managed by the provincial RTVUs, and prepares relevant documents for practice teaching in the compulsory courses. OUC also provides guidelines for directing students' graduation design or graduation project during practice teaching, and trains and certifies RTVU Course Coordinators who are in charge of the graduation project defense.

Evaluation

As discussed, OUC prepares all relevant documents to standardize the teaching processes of the provincial RTVUs and directs them to develop their own rules and documents, following OUC guidelines. OUC also provides *Working Procedures for the Compulsory Courses of OUC*, which stipulates the principles of assessment, instructions for compiling and marking the assessment, duties and responsibilities of the exam teams, development and management of the exam questions, arrangement and organization of each course assessment, and the statistical analysis of course assessment results.

Teaching quality of all courses offered at the provincial RTVUs is regularly audited by OUC using the criteria developed to measure the quality of distance teaching processes and material uses. OUC consults the teachers and students at the provincial RTVUs with regard to the improvement of teaching quality.

Moreover, OUC undertakes comprehensive tracer studies of RTVU graduates in order to measure teaching outcomes and improve the teaching process. These studies collect various data including the enrolment and graduation years, learning centers, program completed, previous work history, current occupation and position, promotion after graduation, technical or research project completed after graduation, published research, awards received, job changes, and further studies. These follow-up studies of DE graduates play a very important role in measuring the quality of teaching performance and learning outcomes. They help OUC determine the needs of students and employers, and adjust the structure of its courses to improve teaching achievement and learning outcomes. OUC has been conducting these nationwide surveys of graduates produced under the Open Education Pilot Project since 2004. OUC has also directed the provincial RTVUs to carry out similar surveys.

Learning Support Services

Learning support services refer to measures taken to assist students to learn continuously, reduce the loneliness of studying at a distance, and achieve learning goals (Zou 2009). In order to support student learning, OUC provides various multimedia materials, course syllabi, formative assessment brochures, exam instructions, and other learning resources. Based on a comprehensive consideration of student needs, especially different learning conditions in the economically developed and less developed districts, OUC offers three kinds of multimedia materials—printed, audio, and video materials—for each compulsory course before every semester begins, and provides the online resources.

In addition to the provision of these multimedia materials and support by phone, email and/or face-to-face mode, OUC uses the online environment to:

- Regularly update learning requirements for each course;
- Provide feedback to students' enquiries;
- Post contact information such as e-mail addresses of the course coordinators and their offices;
- Moderate online forums; and
- Offer pre-examination materials to help students prepare the end-of-semester examination.

OUC provides various services to the teachers who are in charge of learner services at the provincial RTVUs. First, all teachers can access information needed for their teaching and research on the teachers' website, receive advice from OUC staff by phone or e-mail, obtain feedback on the results of their teaching evaluation, and attend training conferences. As mentioned, teachers at the provincial RTVUs are also provided with teaching strategies for each program and course, directions for conducting examinations, course materials, and other resources for their special use.

Moreover, OUC implements training programs for the teachers at both OUC and the provincial RTVUs on how to develop courseware, make good use of Web technologies, apply the theory of DE to their practical work, and conduct research on their teaching. In particular, OUC strongly supports the teachers at OUC and RTVUs to conduct research activities, including academic research, DE research and technology-related research. OUC also organizes seminars and forums for the teachers in order to cultivate an academic atmosphere and encourage staff to be more innovative in their teaching. OUC also provides part-time teachers with orientation programs on the characteristics of DE in order to help them develop understanding of DE in general and acquire effective distance teaching approaches.

Teaching Management

Teaching management is as important as teaching itself in a mega-university such as OUC. OUC manages its teaching at several different levels.

At the whole system level, OUC operates a management mechanism that coordinates planning, implementing, and managing processes at OUC and RTVUs, at provincial, municipal, and prefectural levels, and develops collaborative working relationships. For this purpose, OUC takes responsibility for overall planning, nationwide program setting, developing the overall teaching plan, managing the whole teaching processes, defining unified courses, offering training programs, providing course plans, assessment guidelines and course materials, and monitoring the teaching implementation of the unified courses. In addition, OUC sets the standards and administrative methods for managing the learning centers and testing sites according to the relevant requirements of the MOE, organizes expert meetings to monitor and evaluate its teaching environments, facilities, and teaching staff, and provides support for the teaching activities at the learning centers (Xie, Jiang, Pei, Huang, Wang, and Yan 2010).

OUC manages the information statistics system by providing a planning and implementation scheme for preparing the annual report according to the relevant rules and requirements of the MOE, assigning data collection work to the RTVUs at different levels, and assuring the reliability and validity of data collected.

OUC also manages its human resources in a coordinated manner. It sets position, responsibility, and qualification requirements for teaching staff, chief editors of course materials, and Course Coordinators. OUC evaluates its teaching staff by applying teacher evaluation standards, methods and procedures, and records, analyzes and uses the evaluation results. Furthermore,

OUC establishes training regulations for full- and part-time teachers, develops the training plan and implementation scheme, organizes and carries out various teacher training activities, and summarizes, analyzes, and provides feedback on training effects. With respect to the student information system, OUC is responsible for developing the unified enrolment promotion outline across all RTVUs, publishing enrolment information, and strengthening the management of enrolment information dissemination. It is also responsible for managing the student profile of the whole RTVU system, setting up and implementing guidelines for granting the graduation certificates, reviewing the procedures for graduation, and issuing the graduation certificates according to the relevant procedures.

In managing its teaching resources OUC is responsible for establishing standards for the development of course resources, providing effective channels for course delivery, and evaluating the effective delivery of teaching resources.

OUC manages its intellectual property by being responsible for investigating and analyzing problems with respect to the intellectual property across the whole RTVU system, establishing, elaborating, and standardizing regulations concerning intellectual property and implementing the intellectual property management scheme.

Finally, OUC manages learning assessment by planning, regulating, and organizing training for staff on learner assessment, and managing formative and summative assessment activities.

Teaching and Learning Environment

The teaching and learning environment and teaching tasks in the RTVUs differ, depending on their operational levels—provincial, municipal, and prefectural. In order to be adaptive to these different situations. OUC defines the minimum standards required for the necessary teaching and learning conditions for launching a new learning center at the different levels. This usually includes the basic equipment requirements for classrooms, examination sites, laboratories, libraries, live distance teaching systems, network construction and application, and qualifications of teaching staff.

OUC also monitors teaching conditions at the RTVUs, and allocates special funds for infrastructure improvement to ensure effective teaching and learning. The effect of this investment is to extend the functions of the online education platform, use satellite technology to deliver teaching resources to the provincial RTVUs, improve live distance teaching systems, and construct and share electronic libraries and online testing systems.

OUC operates its own human resources bank as the lead organization of the RTVUs system in order to ensure the smooth implementation of DE. The human resources bank contains personnel information, including work regulations and defined responsibilities for each position, evaluation results of staff performance, and training and support records.

ISSUES AND CHALLENGES IN QUALITY ASSURANCE AT OUC

Need for an Integrated Resources Improvement and Sharing Mechanism

OUC has made significant progress in the development, management, and dissemination of teaching resources in various media formats, applying flexible regulations. However, partly due to the size and complex structure of OUC and the RTVUs, not all of these resources have been adequately updated, shared, or modified to meet the different needs of teachers and students at the different levels of RTVUs. An integrated mechanism is needed to more effectively and efficiently update, modify, and share these resources across the whole OUC and RTVU system.

Need for Continuous Improvement in Distance Teaching and Examination

Although OUC has paid much attention to the design of the distance teaching and examination process, the utilization rate for audio-visual teaching materials is low and the students' assessment results are not satisfactory. This suggests a need to improve training and support for teachers in the use of multimedia materials, and to improve student assessment methods such as through the use of Web-based examinations.

Need for Timely Feedback and Efficient Information System

As discussed above, OUC has built a learning support system that integrates with various learning and tutoring resources, and applies synchronous and asynchronous network technologies to provide flexible tutorial and administrative services nationwide. However, there is still need for more timely feedback from teachers to students, and for more efficient and streamlined procedures for information distribution.

Meeting Challenges of New Paradigms in Education

The introduction of new teaching and learning models in DE has resulted in new and challenging requirements for the teaching and management staff. OUC has realized the urgent need to pay more attention to updating its regulations, strengthening its research capacity, exploring recent teaching and management theories and practices, and extending functions of its current teaching management platform to incorporate new paradigms in education.

Need for Improved Staff Training and Development

Successful distance education depends largely on the quality of teaching staff. OUC engages a large number of part-time tutors, and an effective training and development program is needed to improve their ability to conduct effective tutorials and to enhance their commitment to distance teaching.

LESSONS LEARNED

Comprehensive Quality Management and Total Involvement

OUC has learned that both comprehensive quality management and total involvement of all stakeholders are essential to achieve effective QA in DE and to prevent members of an institution from developing negative attitudes toward DE. This can be achieved only when all divisions, full- and part-time staff, and all levels of operation work harmoniously toward the provision of DE, and when students actively participate in distance learning.

OUC has also learned that efficient and integrated process management is important for QA in DE. QA measures should be applied throughout the entire teaching and learning process, from program/course design to final examination execution and granting the graduation certification, and from enrolment to graduation.

All DE institutions should establish QA systems that align with their philosophy and vision. In the case of OUC, its vision aligns with the strategic goal to ensure teaching quality and cultivate qualified professionals, and is reflected in the DE quality standards and comprehensive quality management regulations and processes.

Focusing on Uniqueness of DE

The OUC experience demonstrates that a QA system for DE should reflect the unique features of distance teaching and management. The OUC's most recent QA system is built on the basis of a Web-based learning environment and a hierarchically operated school environment. It emphasizes the importance of rich distance learning resources and associated learning support, since learners are often engaged in autonomous learning in a Web-based learning environment. It also pays significant attention to the management of teaching and learning processes, support services, materials, and human resources.

CONCLUSION: FUTURE CHALLENGES

OUC's teaching quality management model is unique, in that OUC plans, operates, and manages QA measures for the whole RTVU system while simultaneously collaborating closely with RTVUs at a range of operational levels. By applying its quality management model to its QA system, OUC aims to provide quality education to several million working adults and thus meet the demands for economic development in China. In order to achieve this goal, OUC should strive to:

- Integrate innovative DE models in its QA processes;
- Continuously improve its teaching quality;
- Modify its teaching quality management criteria to be more adaptable to the characteristics of DE;
- Align QA criteria for DE with the national education reform directions;
- Adjust its goal to widen lifelong learning opportunities;
- Improve the quality of teaching resources and services;
- Conduct further research on teaching quality management;
- Ensure the implementation of comprehensive quality management throughout the RTVU system; and
- Move beyond the standardization and nominalization of QA measures with dynamic control.

Standardized and nominalized management is a model of linear thinking, which is suitable for routine management that is not experiencing dramatic changes in the environment, conditions, and factors. Traditional administrative and academic management should transform itself from an emphasis on a linear management model to a more humanistic and service-oriented approach. The focus of QA should be not on what is convenient for the administrator but on what suits those receiving administrative services.

The transformation of our quality management model, from a standardized to a more adaptive and flexible one, is OUC's main challenge. To meet this challenge, further research into service-oriented and innovative quality management is required, with reference to alternative international and regional QA models in distance education.

REFERENCES

Du, R., Yang, T., Yin, S., and Zhang, X. 2009. "Construction of the Criteria System of Teaching Quality of Open and Distance of China," *Open Education Research*, 15 (4): 32–37.

Xie, S., Jiang, X., Pei, H., Huang, P., Wang, Y., and Yan, X. 2010. *Teaching Management Manual*. Beijing, China: CCRTVU Press.

Yang, T. 2008. *Approach on CRTVUs Academic Quality Assurance System*. Beijing, China: CCRTVU Press.

Zou, F. 2009. *Distance Education Guarantee: Learning Support and Strategy*. Beijing, China: CCRTVU Press.

India's Indira Gandhi National Open University

11

Pema Eden Samdup and Rose Nembiakkim

INTRODUCTION

The University Grants Commission, India, established the National Assessment and Accreditation Council (NAAC) in 1994 as an autonomous body to assess and accredit universities and colleges in India. NAAC is triggering a "quality culture" among the various constituents of higher education in India, as well as enhancing the awareness of quality assurance (QA) with all stakeholders (UGC 2007).

As for the open and distance learning (ODL) institutions, the Distance Education Council (DEC), empowered under Statute 28 of the Indira Gandhi National Open University (IGNOU) Act, has primary responsibility in promoting, coordinating, and maintaining quality standards in the ODL system. In 2011, DEC had one national Open University (IGNOU), 13 state open universities (SOUs) and more than two hundred distance education institutes (DEIs) under its purview. DEC faces significant challenges, as most State Open Universities do not wish to abide by its decisions, their argument being that DEC is not really an independent body (DEC is currently housed in IGNOU and the Vice Chancellor of IGNOU is its *ex-officio* Chairman).

Nevertheless, DEC fulfills its mandate to "provide academic guidelines to promote excellence, encourage use of innovative technologies and approaches, enable convergence of all systems and sharing of resources through collaborative networking for access to sustainable education, skill up gradation and training to all," and has also, over the years, initiated several quality assurance parameters (cited from http://www.dec.ac.in/vision.htm). DEC also accredits SOUs and correspondence course institutes across

India. In 2007, DEC conducted an academic review of IGNOU's Master's degree program in English. The committee comprised a minimum of 30 percent of external members. The audit report was not made public, as it was the first time that DEC had evaluated an IGNOU program. Since 2005, DEC has initiated reforms and parameters for offering online and offline courses for ODL institutions.

DEC has developed guidelines for ODL programs, such as:

- Norms and guidelines for management programs;
- Norms and standards for IT education;
- Guidelines for regulating the establishment and operation of ODL institutions in India;
- Norms for ODL; and
- Norms for online programs.

DEC has identified and described seven key areas as prerequisites for ensuring the quality of learning and teaching in online education:

- Organizational structure;
- Planning and development of academic programs/ courseware;
- Design of e-content;
- Student admission;
- Learner support services;
- Learner assessment and evaluation; and
- Technology infrastructure and use.

DEC, being the quality custodian of ODL in India, works alongside other national agencies such as the All India Council of Technical Education (AICTE) and the National Council for Teacher Education (NCTE), and has laid-down norms for regulating Technical and Professional Academic Programs offered by ODL institutions in the country. The National Assessment and Accreditation Council (NAAC), another important national agency that safeguards quality in higher education in India, has worked in collaboration with DEC in preparing parameters for both assessing as well as accrediting open and distance education institutions in the country.

INDIRA GANDHI NATIONAL OPEN UNIVERSITY

Indira Gandhi National Open University (IGNOU), created by an Act of Parliament in 1985, is working toward democratizing education, providing access to quality higher education to all strata of society (including the disadvantaged and the marginalized). As of April 2011, the University serves the educational needs of over two and a half million learners in the country, as well as overseas. IGNOU currently has 21 schools of study and a network of 62 regional centers, and around three thousand learner support centers and 67 overseas centers. IGNOU promotes, coordinates, and regulates the standards of education offered through ODL in the country through DEC (IGNOU 2011). "The QA policies and regulations in IGNOU are in conformity with the QA guidelines determined by the national quality agency, the Distance Education Council of India" (Jung 2005: 82).

IGNOU operates nationally as well as internationally by offering flexibility in terms of entry level qualifications, place, pace and duration of study, as well as modular programs with different exit points. It provides a wide range of cost-effective academic programs at various levels of certification and degrees to highly heterogeneous learner groups. It boasts a rigorous course development mechanism to ensure quality assurance and promotes the use of ICT in teaching and learning.

Under the system of QA, institutions of higher education under consideration abide by the mandatory guidelines, and any academic program, learner support mechanism or other relevant aspect has to pass all the pre-assigned quality parameters. QA thus plays a crucial and critical role in designing the vision and mission statement of an institution of higher learning, as well as in shaping both its outlook and the learner-friendly environment it creates. Where the learners are at a distance, as is the case with IGNOU and other such ODL institutions, QA plays a vital role in the day-to-day activities of the institution.

IGNOU has now been in existence for more than a quarter of a century, but questions related to the credibility of ODL are still raised. ODL in India is still looked upon as a poor or second cousin to the conventional system of education by a vast majority of people. As recently as 2006, doubts over the credibility of ODL programs offered by IGNOU were raised (One India 2006).

Nonetheless, IGNOU is also an important "exporter" of ODL programs, and has also set QA guidelines for exporting programs. First, the credibility of partner institutions (PIs) is established and is reviewed in collaboration with the Indian High Commissions and embassies abroad. IGNOU also approves local tutors and academic counselors, both at home and abroad, based on their curriculum vitae (Jung 2005). These selected counselors are then trained and oriented by the faculty at IGNOU.

QUALITY ASSURANCE AT IGNOU

An effective QA framework for ODL institutions is important, and requires careful and long-term planning and development. Several papers have addressed the need for a QA framework (Gandhe 2009; Prasad & Stella 2006; Rausaria 2002), and QA in ODL has been well-researched by other practitioners from IGNOU as well as other ODL institutions (Jung 2005; Panda 2005; Ramanujam 2000). IGNOU is now a mega-university (Daniel 1996) with a cumulative enrolment of two million, six hundred thousand learners, about three hundred and fifty faculty members and other academic staff, yet does not have a consolidated QA framework in place. Given the nature and scope of this vast institution, some time is likely to elapse before a QA framework for the University in its totality emerges. This, however, does not imply that IGNOU does not have any QA checks in place. To compensate for the lack of an overall QA framework, IGNOU has a rigorous QA mechanism for course development and quality checks in student support services.

The following sections will discuss the QA mechanisms in place at IGNOU with regard to curriculum design and course development, and analyze issues and problems emerging from the QA activities.

QA in Curriculum Design

Since its inception, IGNOU has become well known for the quality of the learning material that it produces. The learning material developed by IGNOU was initially known as Self

Instructional Material (SIM). In keeping with the evolution of ODL methodologies, SIM has now been rechristened Self Learning Material (SLM). In the early years, curriculum design and course development work was undertaken by a course preparation team, but this task has now moved to individual faculty members in the various schools of study.

At IGNOU, New Delhi, there is a clearly defined format for curriculum design. This involves the careful planning of an academic program, after taking into account the following factors:

- Needs analysis: expertise of the faculty, approval of the governing bodies, funding agencies (if any), industry needs, learner needs;
- Needs of the target groups: target population, employment opportunities, economic viability, academic relevance;
- Prospective learner profile: demographic background of the students, motivation, learning skills;
- Preparation of a concept note: need for the academic program, market demand, academic relevance, social relevance, program objectives, target groups, program level and type, medium of study, outline of the concepts, strategies to develop and deliver the program, meeting the objectives of the program;
- Outlining the program structure: context within which the program is envisaged, assessment of needs, developing program aims and objectives, outline of the course; and
- Costing and budgeting: preparation of budget and cost calculations, cost concepts, economies of scale, funding, decision on student fee.

The program proposer is also expected to design and develop the Program Guide in such a manner as to provide all the necessary details and information about the program as well as the University to the distance learner.

The Staff Training and Research Institute of Distance Education (STRIDE) has a well-designed manual for Program and Course Coordinators for curriculum design. Figure 11.1 shows the procedures involved in the Program Approval process (STRIDE 2006: 27).

Figure 11.1   Program approval process at IGNOU

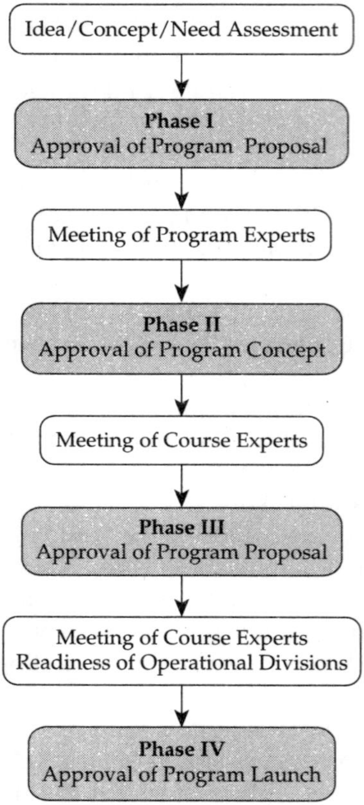

Source: STRIDE (2006, p. 27).

QA in Course Development

Course development is the process whereby quality SLMs are designed, written, printed, and delivered as per the needs of the academic program under consideration. Several courses make up an academic program. For instance, a 64-credit Master's degree in English program at IGNOU comprises eight courses of eight credits each. The course development process is another area where quality checks are stringently maintained. Several crucial factors that affect the course development process include: "the level at which courses must be approved; course production deadlines; varying views of what constitutes respectable materials; shortage of working times; availability of instructional support services; faculty experience with distance education; adaptability of the faculty to the course development activity; interpersonal relationships between members of the course team; and course development models" (STRIDE 2006). The course development process has several stages, as shown in Figure 11.2.

It is the task of the Course Coordinator to play several roles, as the Subject Matter Expert (SME), Unit Writer, Content Editor, Language Editor, Format Editor, Surrogate Learner, Coordinator/ Project Manager, and Quality Assessor. The printing of the Self Learning Material also has a series of procedures to be followed, but this aspect does not necessarily directly affect most faculty

Figure 11.2 Stages involved in course development

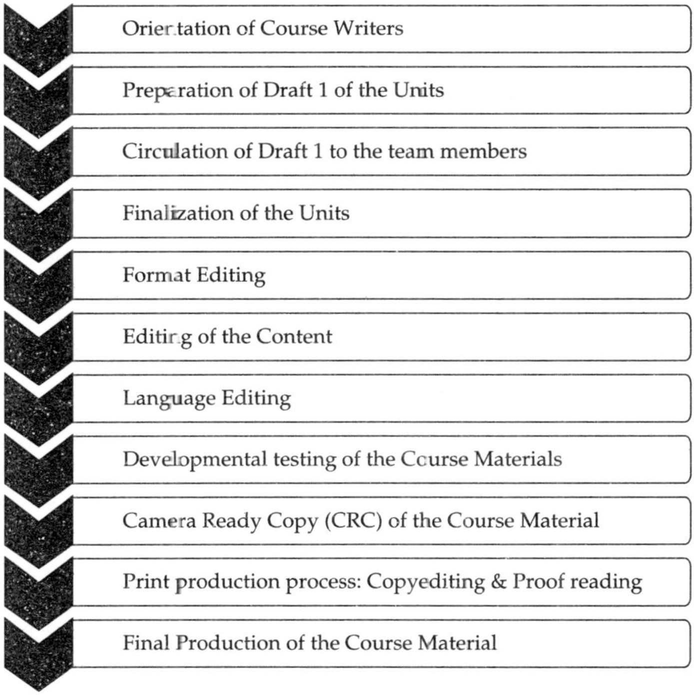

Orientation of Course Writers
Preparation of Draft 1 of the Units
Circulation of Draft 1 to the team members
Finalization of the Units
Format Editing
Editing of the Content
Language Editing
Developmental testing of the Course Materials
Camera Ready Copy (CRC) of the Course Material
Print production process: Copyediting & Proof reading
Final Production of the Course Material

Source: Created by the author based on STRIDE (2006, p. 39).

members. The type of paper to be used, the size, the Camera Ready Copy (CRP) preparation and the layout (page layout/ in-house design of each page) of the SLM and the actual printing process need consideration when analyzing IGNOU's QA mechanisms. In addition, electronic learning material requires a careful assessment of the use of media, the mix of media, and the planning for the audio-video courseware, e-learning materials and interactive media when considering QA.

The manual designed and developed by STRIDE also has a section on course maintenance and the revision of the course in due time. Every Course Coordinator is advised to keep a master copy of the printed SLM and to note all typographical errors as well as any factual errors. Moreover, any development in that area of study should also be noted, so that during the revision process these changes/developments may be added to the SLM. This should be used as a feedback mechanism to collect feedback both formally as well as informally. Sometimes information or sections of material become irrelevant or redundant with the passage of time, and the Course Coordinator should make a note of any such material. Revision is an important aspect in the life of a course, particularly in ODL institutions, so it is necessary and advisable to revise SLM at regular intervals.

When considering revision, a decision needs to be made whether to revise the older material or to withdraw that material totally. A life span for courses thus needs to be in place, especially for programs and courses in constantly evolving professional areas of study. Currently the life span for a course is set at five years, but courses are usually revised within a more realistic time period of ten years.

The manner in which assignments are prepared and the weight given to each need to be taken into careful consideration during the design phase of the courses. Assignments in IGNOU are usually marked by academic counselors hired by the University, who do not technically come under the direct purview of the University. It is imperative that the quality of marking is assured, and the TMAs are monitored on a regular basis.

The academic counselors initially require and receive orientation, as most of them come from conventional colleges or universities and may not be equipped to handle ODL learners. In line with this, the University has recently initiated another quality check measure, whereby a certain percentage of tutor-marked assignments are sent to the headquarters for moderation (evaluation by internal faculty). Further, the delivery of the programs— be it the dispatch of the print material, the registration of learners for the program/course or the dispatch of the electronic media— all need to be streamlined.

IGNOU conducts exams bi-annually (June and December), and the question papers are set by internal as well as external experts. The Student's Evaluation Division (SED) moderates the papers through use of a senior external expert or group of experts and recommends them for printing. Before the grade sheets or cards of the learners are made public, a group of internal faculty members tabulates the results and randomly checks a sample of the evaluated term end answer scripts to ensure that papers have been evaluated properly and malpractice has not taken place.

IGNOU has academic bodies in place that examine each program proposal and monitor the launch of each course, the key bodies for academic matters being the Academic Council and the Planning Board. The Academic Council has established a Research Council and an Academic Council Standing Committee. The relationships between the academic bodies and program development and implementation are shown in Figure 11.3.

Despite all of the above, as mentioned, IGNOU has lacked an overarching QA policy framework. The senior management has taken note and advised that QA measures and good practices are to be consolidated and documented as the IGNOU QA framework.

ISSUES AND PROBLEMS

Even with clear-cut QA measures in curriculum design and course development in place, issues and problems can arise. For example, when a newly recruited faculty member of the University initiates an academic program or develops a course without adequate training or orientation in the course development process, quality issues arise. Further, when there is a rush to announce programs and courses, the QA measures that are in place tend to be ignored. It is imperative that laid-down procedures for curriculum design

Figure 11.3   Academic bodies overlooking QA at IGNOU

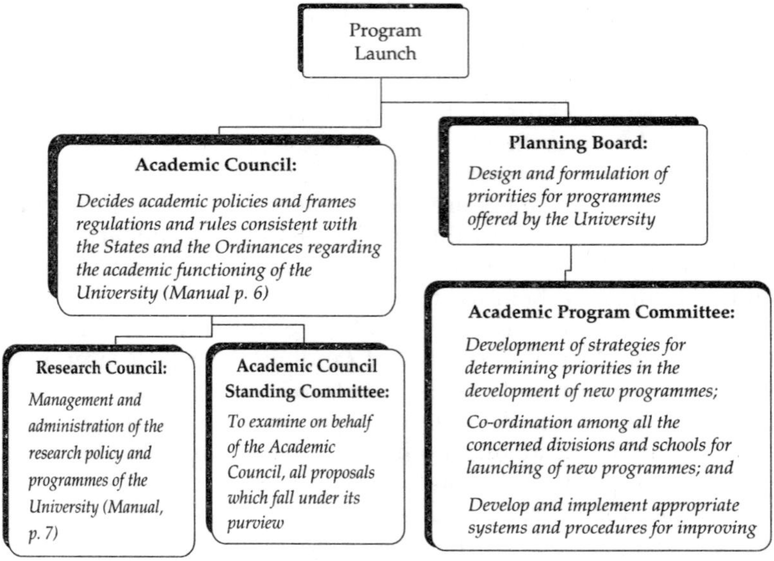

```
                    ┌──────────────┐
                    │   Program    │
                    │   Launch     │
                    └──────────────┘
```

Academic Council:

Decides academic policies and frames regulations and rules consistent with the States and the Ordinances regarding the academic functioning of the University (Manual p. 6)

Planning Board:

Design and formulation of priorities for programmes offered by the University

Academic Program Committee:

Development of strategies for determining priorities in the development of new programmes;

Co-ordination among all the concerned divisions and schools for launching of new programmes; and

Develop and implement appropriate systems and procedures for improving

Research Council:

Management and administration of the research policy and programmes of the University (Manual, p. 7)

Academic Council Standing Committee:

To examine on behalf of the Academic Council, all proposals which fall under its purview

Source: Created by the author based on STRIDE (2006, pp. 6–7).

and course development (as detailed earlier) are followed diligently and thoroughly to ensure the quality of the programs and courses. The Program and Course Coordinators need to follow the quality checks and measures, and involve other faculty members in the Course Development Team. Developing courses independently, an emerging trend, may also compromise quality.

As a quality check, the Distance Education Council (DEC) reviews the course material developed by the schools. Experts from outside the University are invited to review the material, provide feedback on any shortfalls, and recommend changes where needed. This is an effective QA tool, though at times issues have emerged over the subjective nature of some comments.

Problems can also occur when a course is approved by the relevant committees, but course development does not take place for a considerable period of time. When eventually handed over to a faculty member for course development, the school will ask that the course proposal be placed before the School

Board, and that the entire process outlined under curriculum design be followed before the faculty member can undertake course development. The syllabus needs to be redesigned (if considerable time has elapsed since its initiation), a new expert committee has to discuss the syllabus, a fresh course writing committee constituted, and changes or suggestions incorporated. This may be perceived as time-consuming and tedious, but what such a process contributes is the assurance of quality self-learning materials, resulting in a new, dynamic course that meets current needs.

CONCLUSION

A strength of QA at IGNOU is that the various bodies scrutinize every aspect of program and course design and development, and examine each quality activity during the process. The lessons that other institutions can take from the IGNOU experience are as follows:

- New staff and faculty at ODL institutions, especially those recruited from conventional institutions, should be trained to understand the details and nuances of ODL and SLM development before engaging in program and course development.
- Staff should be adequately briefed in the procedures for program and course development and, if possible, an experienced faculty member should be in the team to mentor the newer faculty members.
- Program and course development is a team effort, and in an ODL institution the duties of ODL teachers are typically more complex than those in a conventional system. The human resources team should build the capacity of ODL teachers accordingly.
- In some contexts, ODL has still not been recognized as a quality, alternative mode of education, and is looked upon as a poor cousin to conventional universities. Thus, it is critical for an ODL institution to build and execute strong QA policies for public accountability as well as for self-improvement.

- Well-designed, continuing, and periodic orientation and training of faculty, with regular refresher programs, are necessary for the success of QA in an ODL system.

At IGNOU, QA in curriculum design and course development needs to be more closely linked to the learner support system. Moreover, if a feedback system that gathers stakeholders' opinions and incorporates them into the whole university operations is built into the system, a transparent and strong QA mechanism will emerge. IGNOU needs to revisit its QA practices, consider linking QA in program and course development with the learner support and feedback system, and engage seriously in the creation of a documented QA framework for the whole University system.

REFERENCES

Daniel, J. S. 1996. *Mega-Universities and Knowledge Media: Technology Strategies for Higher Education*, London: Kogan Page.

Gandhe, S. K. 2009. "Quality Assurance in Open and Distance Learning in India," *University News*. New Delhi: Association of Indian Universities.

IGNOU. (2011). "Profile," Indira Gandhi National Open University, New Delhi.

Jung, I. S. 2005. "Quality Assurance Survey of Mega Universities," in McIntosh, C. and Voroglu, Z. (eds), *Lifelong Learning and Distance Higher Education, COL Perspectives on Distance Education Series*, pp. 79–95. Vancouver: Commonwealth of Learning.

One India. 2006. "IGNOU Rebuts Speculation over Degree Credibility," *One India News*, May 15. Available online at http://news.oneindia.in/2006/05/14/ignou-rebuts-speculation-over-degree-credibility-1147644941.html and downloaded on September 28, 2012.

Panda, S. 2005. "Higher Education at a Distance and National Development: Reflections on the Indian Experience," *Distance Education: An International Journal*,26(2): 205–225.

Prasad, V. S. and Stella A. 2006. "Best Practices Benchmarking in Higher Education for Quality Enhancement," *University News*. New Delhi: Association of Indian Universities.

Ramanujam, P. R. 2000. "Quality and Research in Distance Open Learning,", *Indian Journal of Open Learning*, 9(1): 63–72.

Rausaria, R. R. 2002. "Programme Deliver Model based on Use of Interactive Communication Technology and Associated Quality

Assurance Mechanism," *Indian Journal of Open Learning*, 11(3): 319–326.

STRIDE. 2006. *Manual for Programme and Course Coordinators*. Indira Gandhi National Open University, New Delhi.

UGC. 2007. *11th Plan Guidelines for Establishment and Monitoring of the Internal Quality Assurance Cells (IQACs) in Higher Educational Institutions (HEIs)*, *(2007–2012)*, University Grants Commission, India. Available online at http://www.ugc.ac.in/page/XI-Plan-Guidelines.aspx and downloaded on September 28, 2012.

University of the Philippines Open University

12

Patricia B. Arinto

INTRODUCTION

Distance education (DE) in formal degree programs commenced in the Philippines in the late 1970s and early 1980s. According to Natividad (2001), these pre-1990s DE programs were generally "small-scale operations" with low enrolments. Moreover, few people understood DE, and some institutions claiming to offer DE programs were actually delivering extension programs in which lecturers met students at a location off-campus during weekends. It was not until the 1990s that DE began to gain recognition as a legitimate mode of delivering higher education in the Philippines and established universities began to offer DE programs. The Philippine Women's University began offering lessons via television for its DE mode Master of Arts in Education program, and the Polytechnic University of the Philippines and Visayas State College of Agriculture formally established themselves as "open universities" in 1990 and 1997, respectively (Natividad 2001). The University of the Philippines – Open University (UPOU) was established in 1995 as the fifth constituent university of the University of the Philippines System, the national university.

UPOU's mandate is to offer formal and nonformal programs via DE mode, and thereby improve access to a UP education, which is widely considered to be the best in the country. Its first program was the Diploma in Science Teaching, first offered in the University of the Philippines at Los Baños in 1987. Today, UPOU is the most comprehensive DE provider in the Philippines, with three undergraduate programs and 19 graduate degree programs across three Faculties—the Faculty of Education, the Faculty of Information and Communication Studies, and the Faculty of Management and Development Studies. It also offers several nonformal courses. The average total enrolment per semester is about 2,500 students. The low enrolment, relative to enrolments in other open universities in Asia, may be attributed to the following most program offerings are at the graduate-level; there are restrictive admission policies for the undergraduate programs (students must take and pass either the UP College Admission Test or the UPOU Undergraduate Admission Test and, in the case of the BA in Multimedia Studies, have at least five years of work experience in a field related to multimedia design); and tuition fees are relatively high compared to fees charged by other state colleges and universities.

With the increased interest in DE among tertiary level institutions in the 1990s, the Philippine Commission on Higher Education (CHED) began to establish policies to prevent the offering of substandard programs by DE mode (Natividad 2001). Under the Higher Education Modernization Act (Republic Act 8292), CHED supervises all public and private higher education institutions (HEIs) in the Philippines. Most state colleges and universities undergo voluntary accreditation by the Accrediting Association of Chartered Colleges and Universities of the Philippines (AACCUP). Private HEIs are required by CHED to be certified by the Federation of Accrediting Agencies of the Philippines (FAAP), which includes the Philippine Accrediting Association of Schools, Colleges and Universities (PAASCU) and the Philippine Association of Colleges and Universities' Commission on Accreditation (PACUCOA). There is no accrediting body solely for DE institutions and programs.

CHED issued its first set of guidelines for open learning and DE in December 2000. Besides requiring institutions intending to offer DE programs to obtain an initial permit, the guidelines provided for the formation of a Technical Committee to evaluate proposed DE programs. In 2005, CHED issued Memorandum Order No. 27 titled "Policies and Guidelines on Distance Education." This stipulates that only graduate-level programs with Level III accreditation can be offered by DE mode. Level III accreditation is granted to institutions that have a "high standard of instruction" and a "strong faculty development tradition" and at least two of the following:

- Highly visible community extension program;
- Highly creditable performance of its graduates in licensure exams;

- Working consortia or linkages with other schools/agencies; and
- Extensive and functional library and other learning resource facilities (PAASCU n.d.).

The CHED memorandum order defines a DE program as one where at least 25 percent of all courses are offered via distance mode. An HEI seeking to offer a DE program is required to obtain authorization from CHED to offer the program, and then undergo periodic monitoring and evaluation by the CHED Technical Panel on Distance Education. The following evaluation criteria are applied to HEIs seeking government authority to offer DE programs and/or government recognition of its DE programs:

- Institutional qualification: The HEI must have Level III accreditation for the program it intends to offer by DE mode, or should be a recognized CHED Center of Excellence in that program.
- Institutional management and commitment: The HEI must have a DE unit with a mission and policy statement congruent with the HEI's overall mission, a financial or budgetary allocation for DE operations, an organizational structure and procedures for managing the DE unit, a DE unit head or manager with the appropriate qualifications, and a plan for continuing self-evaluation for program improvement.
- Curriculum development and approval: There should be a complete program curriculum and detailed syllabi defining appropriate learning outcomes and assessment methods for each course.
- Instructional materials development: Self-instructional learning packages should have been developed by a team of qualified subject matter specialists, instructional designers, and production design specialists. In addition, these learning packages must have undergone testing, and they must be compliant with copyright laws. Moreover, the HEI must have defined procedures or guidelines and policies for periodic review and updating of the learning packages.
- Delivery mode/strategies: There should be a core of qualified teaching faculty; policies and procedures for course

delivery; mechanisms for enabling interaction between teachers and students and among students; a validated system of student assessment and evaluation; and resources for learning such as libraries and learning center facilities.

• Student support services: Students should be given complete and clear program information, including registration advice; study and technology skills training; access to grievance procedures; regular feedback on academic progress; and clear admission and retention policies and procedures.

As a constituent unit of the UP System, which has autonomous status, UPOU is not subject to monitoring and evaluation by CHED, and it is not obliged to undergo CHED accreditation of its programs. However, UPOU's quality assurance (QA) standards and procedures are congruent with those specified by CHED in Memorandum Order No. 27, which was formulated by the Technical Panel on Distance Education that included three DE experts from UPOU. In addition, UPOU's programs comply with internal QA policies and procedures governing all UP units, in keeping with UP's status as the premier university of the Philippines. This chapter examines the QA processes implemented by UPOU, particularly in online distance course development and course delivery. Lessons learned and reflections on future directions of QA in DE in a digital age are presented.

QUALITY ASSURANCE AT UPOU

The DE model implemented at UPOU is one where students engage in guided independent study of mostly text-based course modules, and online tutorials conducted through a Moodle-based virtual learning environment (VLE). All course examinations are proctored, with students taking examinations at the UPOU learning centers where they are registered, or at an accredited UPOU testing center. UPOU has a network of 10 learning centers and 19 testing centers in the Philippines and abroad. UPOU testing centers abroad are mostly Philippine consulates and offices of organizations with which UPOU has established partnerships for this purpose.

Online tutorials, using an open source learning management system, were introduced in 2001 for students who were unable to attend face-to-face monthly tutorial sessions. By 2004 all courses had shifted to online tutorials as UPOU students became more widely dispersed across the Philippine archipelago and other parts of the world, and as the internet became more accessible to teachers and students alike. In 2007, UPOU shifted to a Moodle-based platform which allows for the creation of courses featuring digital resources, such as online journal articles and websites, and online activities, such as discussion forums, chat rooms, and quizzes (Blin and Munro 2008).

QA in Program Proposal Development

QA in course and program delivery at UPOU is distributed across several units within the University. The Office of the Vice-Chancellor for Academic Affairs (OVCAA) oversees QA in curriculum development. Course proposals undergo a series of reviews at various levels, beginning with the Curriculum Committee of the proponent Faculty, followed by the University Curriculum Committee, which is chaired by the VCAA.

The institution of a new program follows the process shown in Figure 12.1. The VCAA brings the program proposal to the Academic Affairs Committee (AAC), which is composed of the Vice-Chancellors for Academic Affairs of the seven constituent

Figure 12.1 Flow chart for academic proposals

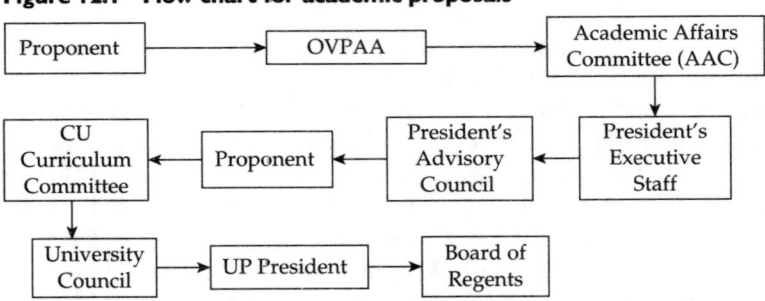

Source: Office of the Vice-President for Academic Affairs (OVPAA). (2011). Annex 3. Flow Chart for Academic Proposals. University of the Philippines. Quezon City.

universities (CUs) of the UP System and headed by the Vice-President for Academic Affairs. From the AAC, the proposal is submitted to the President's Advisory Council, composed of the UP President and the Chancellors of the CUs. The proposal is then submitted for approval by the UPOU University Council (UC) and, finally, by the UP Board of Regents.

A program proposal is evaluated in terms of curricular soundness and operational viability or feasibility.

- Curricular soundness refers to whether the program is anchored in a recognized discipline or field of study, and whether this discipline or field of study is within the purview of the faculty proposing the program; the program goals are appropriate for the level specified; the curriculum structure (including the course offerings), modes of teaching, and modes of assessment are congruent with program and course goals (specified in individual course proposals); and program policies (e.g., regarding the admission and graduation of students) are consistent with existing University policies.
- Operational viability is determined in terms of whether the proposed program addresses a recognized gap in education provision; there is a significant demand for the program from the target population; the program is in line with the mission and goals of the faculty proposing it and there is a sufficient number of qualified faculty members who can teach the courses; the cost of program development and delivery is reasonable; and the fiscal resources needed for this purpose are available.

In addition, at the UP System level, a key consideration is ensuring that the proposed program does not duplicate existing offerings in other UP units.

QA in Course Development

Once a program is instituted, course development begins. In print-based course development, which was implemented at UPOU in its first ten years of operation, QA was the responsibility of Course Teams working under the supervision of the Office of Academic Support and Instructional Services (OASIS) in cooperation with the Faculty offering the program. Using the Course Team approach, which was popularized by the UK Open University, a course package was systematically designed and developed by course authors and reviewers working closely with an instructional designer, editor, and other academic support staff (Haughey 2010; Naidu 2007). Most of the course authors and reviewers, who were recommended by the Deans on the basis of their subject matter expertise, were full-time faculty of the bigger UP constituent universities. UPOU's full-time faculty members, trained in instructional design, acted as the instructional designers. Editing was outsourced to freelance editors.

Most of the printed course materials that were developed following the process described above are still being used in UPOU. However, faculty members who deliver the courses are now being encouraged to supplement these printed modules with Web resources, following a resource-based course design model (Calvert 2005; Naidu 2007). Thus, many course packages now consist of compilations of open educational resources (OERs), including articles from open access online journals and rich media materials, such as video recordings of lectures on YouTube and other Web channels. Faculty may also propose the development of original video and audio materials as part of the course packages. This process is overseen by the Multimedia Center, which appoints a multimedia designer and production personnel.

QA in Course Delivery

In the course delivery phase, QA is overseen by the Dean. Each course is handled by a faculty-in-charge (FIC) who is assigned by the Dean on the basis of his/her expertise in the subject matter. Because UPOU has only 20 full-time faculty—the offshoot of UPOU founding officials' decision to rely on the faculty of the more established UP units in order to forestall doubts about the quality of educational provision by the new UP unit (Arinto 2007), the majority of the FICs are affiliate faculty from the other UP campuses. There are also part-time adjunct lecturers, who are experts connected with organizations outside of the UP System. New FICs undergo workshops in DE course delivery, in particular sessions on how to use the VLE. Tutors may be appointed to assist

FICs in handling large classes, and they too undergo training in DE course delivery.

To measure quality in course delivery, course evaluation questionnaires are administered to students at the end of the semester. The evaluation form includes questions about various aspects of course delivery, including the way the course is handled by the FIC and tutor and the course materials used, and aspects of course design, including assessment of learning (i.e. assignments and examinations). The course evaluation results are used in deciding whether the course materials should be revised or redeveloped, as well as in deciding whether to continue to assign the course to the same FIC.

ISSUES AND POSSIBLE SOLUTIONS

Faculty Preparedness for Online Teaching

QA-related issues during course delivery tend to be readily apparent, perhaps because their repercussions are felt more immediately. These issues include the following:

- Many FICs fail to prepare their course sites or virtual classrooms for student use before the start of the semester. Thus, students find these course sites bare, without basic resources such as a course guide to give students an overview of the course, an introductory message or instructions from the FIC, and an introductory forum where students might at least introduce themselves.
- Many of the functionalities of the online learning platform are not sufficiently utilized to engage learners, enhance learning, and foster collaboration and knowledge building. For example, few FICs have tried the online quiz tool, the chat feature, and the blog and wiki features.
- The most used functionality or feature of the VLE is the discussion forum. However, there is minimal and sometimes no teacher moderation of or participation in these forums beyond setting the forum topics.
- Timely feedback on student work is not given in many cases, and there is a high incidence of delayed submission of grades.

All of these issues point to a seeming lack of preparedness for online teaching on the part of many FICs. To address the problem, the Faculties, in cooperation with OASIS and the TechSupport Desk which is currently lodged at the Multimedia Center, conduct workshops every semester on how to use the online learning platform. These workshops are run by a team of regular faculty members who are given the special additional assignment of providing technology support to FICs. They conduct individual tutorials with FICs who request such sessions, as well as run the Moodle workshops.

Training in Online Course Design

In strategic planning meetings organized by the OVCAA, it has been noted that one reason for the persistence of the issues cited above is that the Moodle tutorials are designed only for new FICs, who attend them once, just before they commence their first semester of teaching at UPOU. While FICs may request special sessions thereafter, few seek such sessions and many FICs have only a rudimentary understanding of how the VLE works, and are thus unable to use it effectively. A basic workshop on how to use Moodle is clearly inadequate for learning how to design online courses well, even for FICs who are "fast learners."

Furthermore, to teach effectively with technology, faculty members need to develop not only the requisite technology skills (i.e., skills in the use of particular hardware and software) but also a deep understanding of how using particular technologies might change the dynamics of teaching and learning (Arinto 2008). Using a VLE like Moodle, especially in a DE context (i.e., where teachers and students do not meet face-to-face in a physical classroom), requires teachers to think carefully about how they should chunk and sequence course content, what types of learning resources to make available, and what mix of individual and collaborative activities to include in order to foster deep and active learning. In short, using a VLE requires teachers to rethink old pedagogies. It is therefore necessary to design and implement a faculty development program to address this aspect of becoming an effective online educator.

One step in this direction that UPOU has taken in the past year is to preface the basic Moodle tutorials with a workshop on rapid instructional design (ID). The rapid ID workshop highlights the online teacher's role as course designer, which involves developing, updating and/or customizing course content, and designing learning activities (Anderson 2008). The workshop teaches FICs how to quickly develop or update content by selecting and repurposing learning resources from the Web, and how to rapidly design learning activities that use Web tools to enable learners to individually and collaboratively explore the learning content and achieve the target learning outcomes.

Updating Course Materials

While it is intended primarily to improve the quality of online course delivery by UPOU faculty, the rapid ID workshop is also a way of addressing a key QA issue in course development, namely, the problem of outdated course materials. The need to revise or redevelop course packages is well recognized in UPOU. However, the actual process of course redevelopment poses significant challenges. It requires, for example, organizing Course Teams, which in turn requires the reorganization of faculty workloads and reallocation of time and other resources. Encouraging and equipping FICs to update course packages through the rapid ID workshop is a quick solution to the problem. It is also a solution that is consistent with emerging models of online course design.

In the industrialized print-based model of course development, standardization of products and outputs was ensured through automation, division of labor among specialists working as a team, and centralized control. This, according to Peters (1998: 117, quoted in Burge and Polec 2008), resulted in the creation of high-quality learning material that is "pedagogically suitable, reflects the latest levels of research, and is presented particularly effectively." In contrast, in a postmodern, postindustrial, and digital era, the development of new courses is undertaken by what Power (2007: 65) refers to as "a reduced version... of the Course Team approach," where one faculty member both designs and delivers a course asynchronously online. The approach taken to course design appears more and more to be "ad hoc" rather than "planned" or "centralized" (Davis 2001: 10) and the design process is much more fluid, taking place not just before but throughout course delivery, which makes it difficult to observe systematically (Arinto 2010).

Institutionalizing a Formal Review Process

The blurring of the boundaries between course development and course delivery in Web-based DE is well documented in the literature (see, for example, Abrioux 2001). In relation to QA, it requires careful consideration of how and when to monitor and assess both the process and outcomes of course design. Presently at UPOU there is no formal review process in place for the evaluation of a course package that has been updated by an FIC using the rapid ID approach *before* the package is used in a course. For example, while FICs are required to develop course guides, there is no formal review of the course guides prior to deployment to assess whether they provide adequate guidance to students and whether key course components, such as course assignments, are well designed. With regard to quality criteria, there is a need to rearticulate design standards for ensuring course accessibility for all students (including student readiness for online learning); the use of appropriate types and levels of interactivity; and the selection and presentation of relevant, credible, and appropriate content while complying with norms governing the use of proprietorial and open educational resources.

CONCLUSION: LESSONS LEARNED

This section concludes this chapter with a synthesis of lessons learned from UPOU's experience in Web-based course design, development, and delivery thus far. DE institutions that are embarking on an online mode of delivery might find these lessons instructive.

First, design and implement a holistic and sustained program for faculty development in online course design and development.

Faculty development in DE course design and delivery should not be limited to the first few years of the establishment of a DE institution. Courses need to be updated all the time, particularly in this era of rapid knowledge production. Moreover, Web-based tools for updating and (re)designing courses or course components are increasingly available to institutions and individual faculty. It is therefore necessary to systematically equip *all* faculty members with the pedagogical frameworks and technical skills for Web-based course design.

Mishra and Koehler (2006: 1030), among others, have pointed out that "[t]he addition of a new technology is not the same as adding another module to a course. It often raises fundamental questions about content and pedagogy that can overwhelm even experienced instructors." Thus, the approach to faculty development in online teaching should be holistic and long-term. It should provide for just-in-time learning, and should therefore be implemented throughout the year and in ways that allow faculty to participate wherever they are. It should also be a program that models the principles of effective online teaching and learning. These attributes point to an online faculty development program that is perhaps less like a course that positions faculty members as students, which is inconsistent with their self-identity as faculty, than an online community of scholars and professionals learning from each other's best practices and engaging in collaborative critical reflection on their work as distance educators.

Second, recognize that modes of practice are changing and adopt new, more appropriate structures and processes in support of QA.

Many DE scholars have written about how online technologies are transforming older models. Among the key changes noted by Calvert (2005: 231) is the increasing emphasis on the design of online learning environments that are "rich in resources" and "rich in social collaboration and interaction." For Tait (2010: x-xi), the shift to online DE has led to the displacement of print as the dominant resource by multimedia resources; the use of resources that are available on the Web in place of production of resources by institutions; the expansion of the mediating role of teachers to include helping learners "to make sense of the wealth of resources which they can, with guidance, find themselves" and increased opportunities for learners to interact and collaborate, which in turn challenge teachers to organize and manage online learning experiences. All of these changes call for corresponding changes in the way that institutions manage and support course development and delivery. More specifically, support services for course development and delivery need to be reconceptualized.

A print-based mode of DE is supported by print production staff, including editors and book designers, and by publication protocols such as those for the review and editing of manuscripts. Online distance educators, on the other hand, need the support of instructional technologists and multimedia designers, among others. For small institutions like UPOU, where human and other resources are limited, meeting this need cannot be a simple case of adding a new unit and hiring new staff, while retaining old units and keeping the current staff where they are. Some reorganization is inevitable, and the institution's leaders need to muster the political will necessary to implement it.

Online distance faculty also need tools and resources to guide practice, such as frameworks, models, and templates. Such tools and resources include quality standards, guidelines, and procedures that faculty can use as reference points for their work. For the institution as a whole, these quality standards and systems are an important means for gauging success in online DE.

REFERENCES

Abrioux, D. 2001. "Guest Editorial," *International Review of Research in Open and Distance Learning*, 1(2). Available online at http://www.irrodl. org/index.php/irrodl/issue/view/10 and accessed on September 7, 2012.

Anderson, T. 2008. "Teaching in an Online Learning Context," in T. Anderson (ed.), *The Theory and Practice of Online Learning*, pp. 343–365. Athabasca University Press.

Arinto, P.B. 2010. "Distance Education Course Design in a Digital Age: Issues and Challenges," Paper presented at the 24th Asian Association of Open Universities Conference, Hanoi, Vietnam, 26–28 October.

———. 2007. "Going the Distance: Towards a New Professionalism for full-time Distance Education Faculty at the University of the Philippines," *International Review of Research in Open and Distance Learning*, 8(3). Available online at http://www.irrodl.org/index. php/irrodl/article/view/409/941 and accessed on September 7, 2012.

———. 2008. "What and how Teachers Know when Teaching with Technology: The Perspectives of Four Filipino Teacher Educators,"

Unpublished EdD Institution-Focused Study Report. Institute of Education, University of London.

Blin, F., and Munro, M. 2008. "Why hasn't Technology Disrupted Academics' Teaching Practices? Understanding Resistance to Change through the lens of Activity Theory," *Computers & Education,* 50(2): 475–490.

Burge, E.J , and Polec, J. 2008. "Transforming Learning and Teaching in Practice: Where Change and Consistency Interact," in T. Evans, M. Haughey, and D. Murphy (eds), *International Handbook of Distance Education,* pp. 237–258. Bradford: Emerald Group Publishing Ltd.

Calvert, J. 2005. "Distance Education at the Crossroads," *Distance Education,* 26(2): 227–238.

Davis, A. 2001. "Athabasca University: Conversion from Traditional Distance Education to Online Courses, Programs and Services," *International Review of Research in Open and Distance Learning,* 1(2). Available online at http://www.irrodl.org/index.php/irrodl/issue/view/10 and accessed on September 7, 2012.

Haughey, M. 2010. " Teaching and Learning in Distance Education before the Digital Age," in M.F. Cleveland-Innes, and D.R. Garrison (eds), *An Introduction to Distance Education: Understanding Teaching and Learning in a New Era,* pp. 46–63. New York and London: Routledge.

Mishra, P , and Koehler, M. 2006. "Technological Pedagogical Content Knowledge: A New Framework for Teacher Knowledge," *Teachers College Record,* 108(6): 1017–1054.

Naidu, S. 2007. "Instructional Designs for Optimal Learning," in M.G. Moore (ed.), *Handbook of Distance Education.* New York and London: Routledge.

Natividad, J. 2001. "Philippines," in O. Jegede, and G. Shive (eds), *Open and Distance Education in the Asia Pacific Region,* pp. 205–219, Open University of Hong Kong Press.

PAASCU. (n.d.). Philippine Accrediting Association of Schools, Colleges and Universities website. Available online at http://www.paascu. org.ph/home2010/ and accessed on September 7, 2012.

Power, M. 2007. "From Distance Education to e-Learning: A Multiple Case Study of Instructional Design Problems," *E-Learning,* 4(1): 64–78.

Tait, A. 2010. "Foreword," in M.F. Cleveland-Innes, and D.R. Garrison (eds), *An Introduction to Distance Education: Understanding Teaching and Learning in a New Era,* pp. IX–XI. New York and London: Routledge.

University of the Philippines Office of the Vice-President for Academic Affairs (UP OVPAA). 2011. *Annex 3. Flow Chart for Academic Proposals.* Quezon City, Philippines: University of the Philippines.

PART 4

Assuring Quality of Learning Support and Assessment

Malaysia's Wawasan Open University

13

Tat Meng Wong and Teik Kooi Liew

INTRODUCTION

In Malaysia, all higher education institutions (HEIs) require prior approval from the Ministry of Higher Education (MOHE) to operate. All programs offered by HEIs must obtain provisional accreditation by the Malaysian Qualification Agency and approval by MOHE before student recruitment. Provisionally accredited programs must undertake full accreditation before the first intake achieves graduation.

The Malaysian Qualifications Agency (MQA) is responsible for assuring the quality of all public and private HEIs, and functions through:

- Administration of the Malaysian Qualifications Framework (MQF) and its related Malaysian Qualifications Register, which lists all qualifications accredited by the MQA. The MQF provides the instrument for classifying qualifications, based on level, number of credits, and learning outcome of the study program. Clear specifications and criteria are laid-down for each program level.
- The implementation of an institutional developmental schema for all Malaysian HEIs whereby each institution progresses through three stages of development toward the achievement of *Self Accrediting Institution* status. These are:
  - o Stage 1: Program accreditation stage, where all new programs are required to be *provisionally accredited* before student recruitment and *fully accredited* prior to the graduation of the first batch of students.
  - o Stage 2: Institutional Audit (IA) stage where the HEI, having completed a few cycles of program accreditation,

applies to the MQA to undertake an IA. Institutions achieving excellent ratings in the IA may be invited by the minister to undertake the Stage 3 audit.

o Stage 3: Self Accrediting Institution Certification (SAIC). Institutions achieving the SAIC status need not have their academic programs undergo further MQA accreditation.

For both program accreditation and institutional audit, the MQA currently applies an outcomes-based approach and identifies nine areas for evaluation, with emphasis on self-review accompanied by a demonstration of a continuous monitoring and improvement culture. This reflects a shift in approach from quality assurance (QA) to quality enhancement (QE), with specific focus on encouraging the development of internally designed QA mechanisms. The nine areas for evaluation are summarized in Table 13.1, which lists the number of criteria applied under the "benchmark" and "enhanced" standards for program accreditation and institutional audit.

To guide HEIs in their document preparation, two comprehensive guides, *The Code of Practice for Program Accreditation* (MQA 2008) and *The Code of Practice for Institution Audit* (MQA 2009) were published. Both guides were designed primarily to evaluate conventional face-to-face (F2F) institutions that form more than 90 percent of the HEIs in the country. The *Guidelines to Good Practices: Open and Distance Learning* has recently been released (MQA 2011).

In 2010, the MQA (under ministerial instructions) conducted Academic Performance Audits (APA) on all HEIs in the country (55 conventional universities and colleges, and three ODL institutions). This massive exercise represents a step toward improving the QA systems and practice of HEIs in Malaysia. Previously, no HEI in Malaysia had undertaken a comprehensive institutional audit of its academic processes. All major ODL institutions in Malaysia, namely Open University Malaysia, Wawasan Open University, and the Asia e University, undertook the same APA exercise, though an instrument for the evaluation of ODL systems was not available. Arising from the exercise, four long-standing local public universities and four branch campuses of foreign universities were awarded the SAIC status. The report for Wawasan Open University contained 27 commendations,

Table 13.1 **Nine areas of evaluation for program accreditation and institutional audit**

Areas	*Code of Practice for Program Accreditation (COPPA) (Number of standards)*		*Code of Practice for Institutional Audit (COPIA) (Number of standards)*	
	Benchmark	*Enhanced*	*Benchmark*	*Enhanced*
Area1: Vision, Mission, Educational Goals and Learning Outcomes	7	4	13	5
Area 2: Curriculum Design and Development	19	11	18	9
Area 3: Assessment of Students	11	5	10	5
Area 4: Student Selection and Support Services	20	13	22	12
Area 5: Academic Staff	11	4	12	3
Area 6: Educational Resources	12	10	13	8
Area 7: Program Monitoring and Review	5	4	7	3
Area 8: Leadership, Governance and Administration	11	6	10	9
Area 9: Continual Quality Improvement	3	2	3	1
Total number of standards	**99**	**59**	**108**	**55**

Sources: Malaysians Qualifications Agency. (2009). Code of Practice for Institution Audit (2nd edition): pp. 22–49 and Malaysians Qualifications Agency. (2008). Code of Practice for Programme Accreditation: pp. 7–27.

12 affirmations and only two recommendations across all nine areas evaluated.

Currently, programs offered by all open universities are subjected to MQA's program accreditation without the benefit of an instrument designed for ODL systems. However, since August 2011, the *Guidelines to Good Practices: Open and Distance Learning* have become available to ODL providers.

WAWASAN OPEN UNIVERSITY

Wawasan Open University (WOU), a private not-for-profit HEI, was established in August 2006 with the main objective of enhancing access to industry-relevant quality higher education for working adults in Malaysia, irrespective of race, gender, and socio-economic background. Its vision, "We aspire to be a vibrant learning community that inspires learning, supports innovation and nurtures all-round personal growth," is reflected by its socially inclusive mission: "We commit ourselves to the expansion of opportunities in higher education and to teaching excellence aimed at increasing the level of knowledge and scholarship among all Malaysians."

WOU delivers its courses and programs through a technology-enhanced open distance learning (TEODL) mode. Sponsored by the Wawasan Education Foundation, WOU offers accessible, flexible, and affordable education to the adult community in support of lifelong learning. WOU had its first student intake in January 2007. By July 2011, more than eight thousand, three hundred students had experienced its flexible TEODL education delivery, with three thousand, seven hundred and eighty seven students actively enrolled in courses leading to the award of 38 named qualifications at the sub-degree, degree, and master level. Student age varies from 21 to 72 (72.7 percent within the 21–35 years age group), and the gender balance is 52.5 percent males to 47.5 percent females. Eighty-seven percent are taking undergraduate programs, while 13 percent are following post-graduate programs offered through the School of Business & Administration, School of Science & Technology, School of Foundation & Liberal Studies and the School of Education, Languages and Communication. The first cohort of 38 MBA graduates was produced in 2010. In the second convocation ceremony held in November 2011, 70 pioneer undergraduates students were conferred their first degree awards.

The WOU education model: Recognizing the diversity of their background, WOU admits students to study in the University rather than for specific programs. The Open Entry Admission System (OEAS) admits mature students (≥ 21 for undergraduate and ≥ 35 for postgraduate) with minimal academic qualifications to study provided they meet other criteria. The OEAS makes WOU's programs more accessible in comparison with conventional universities. Once admitted, WOU students also enjoy great flexibility in their choice of study programs as well as their course load each semester. Fees are payable based on the number of courses (subjects) enrolled each semester.

Programs: Each named academic program consists of a set of prescribed courses of 3, 5 (semester long) or 10 credits (year-long). The total number of credits in each program complies fully with MQF specifications. Credits are earned on the successful completion of each course. There is no maximum time limit for program completion, and students can readily change their intended program. They can also take a break from their studies for one or more semesters. In support of lifelong learning, WOU offers its students a multi-entry and multi-exit progression pathway (Figure 13.1). Multi-entry is offered through advanced standing and credit transfer for credits earned from other accredited institutions, while multi-exit is offered through the award of Graduate Certificates and Diplomas that are seamlessly articulated to their related degree programs.

At WOU, a student can exit after a single course, as the university believes that, if a working adult can benefit from the learning experience of taking one course and in so doing increases his/her productivity, then the University will have contributed toward enriching the community.

Figure 13.1 WOU's multi-entry and multi-exit progression pathways

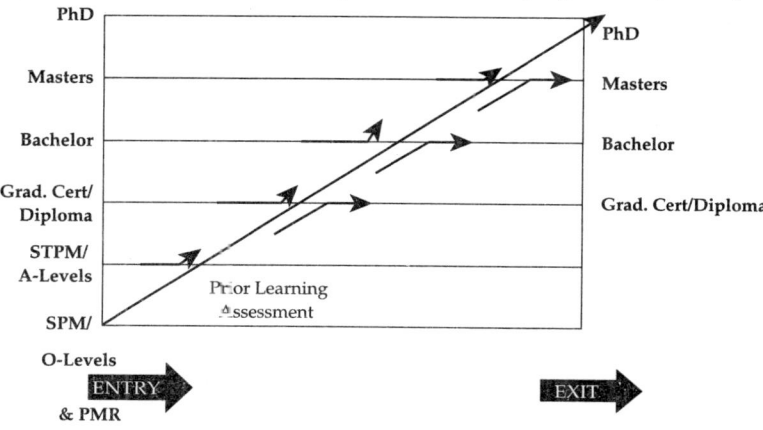

Source: Wong (2011).

Courses: The course delivery model consists of three main components: *(a)* the provision of quality course materials; *(b)* the provision of quality learner support; and *(c)* the application of a rigorous assessment strategy and exit standards.

1. Course materials: The academic content of courses is delivered through pedagogically designed self-learning course materials (including texts books where appropriate) provided via print or CD-ROM. A comprehensive learning guide helps students pace their learning to achieve optimal effectiveness. Pacing is further reinforced through a system of tutorials and scheduled submission of assignments, which also serves formative and summative evaluative functions. The course material's design enables students to engage in learning activities at any time and from anywhere to suit individual learning styles and needs. Supplementary learning resources are provided through an online learning environment.

2. Learner support: WOU provides the following learner support to its students:

   • Tutor support: Tutors are part-time academics with relevant subject expertise. They support the students' learning through 5 x two-hour long F2F tutorials offered at five-week intervals at a local Learning Center each semester and telephone consultation at appointed periods twice a week.

   • Online learning support is provided through a Moodle-based Learning Management System (*WawasanLearn*) that enables students to access additional supplementary materials, link to relevant websites, engage in forum discussions with members of their own tutorial groups or course mates all over the country, as well as consultation with tutors and Course Coordinators on a 24 x 7 basis.

   • Access to the University's extensive digital library resources on a 24 x 7 basis from anywhere with internet connections.

   • The staff and facilities at the seven regional offices (ROs) located throughout the country. All ROs are equipped with tutorial rooms, computer labs, libraries, and free access computer terminals.

3. Assessment: The mastery of learning outcomes of courses is evaluated via an assessment strategy consisting of a number of assignments and a final examination. For each course, the student is required to submit a number of Tutor Marked Assignments (typically three for a 3 or 5 credit course at the lower and middle level and two for higher level and postgraduate courses). These summative assignments contribute to 50 percent (or 40 percent) of the total assessment, the other 50 percent (or 60 percent) being a three-hour proctored written examination. To pass a course, a student is required to pass both components. The requirement to pass the exam component reflects a rigor that surpasses what is required by the MQA and practiced by most Malaysian HEIs.

Programs offered at WOU are market-driven, utilitarian in character, and support the nation's desire to be an active player in the knowledge economy through the production of graduates with high levels of knowledge and skills to enhance the workforce. WOU leverages on advances in technology to enhance student access, increase delivery effectiveness as well as optimize internal process efficiencies, and aims to become a key player in the country's lifelong learning landscape.

This chapter discusses the QA systems and processes that WOU has established to ensure that its students have a quality learning experience and that its graduates are comparable to the best that the country's tertiary institutions can produce.

QUALITY ASSURANCE AT WOU

WOU is committed to providing its students with a high-quality learning experience that leads to the production of quality graduates. This is achieved through the development of a quality culture among its stakeholders, particularly staff and students, that is underpinned by a robust QA system covering all aspects of its operations. QA at WOU is premised on the following:

- Compliance with the criteria and standards laid down externally by the MQA;

- Benchmarking its academic processes against international good practice for ODL;
- Extensive review and intervention by external peers;
- Clearly documented standard operating procedures (SOP) for all core activities;
- Regular extensive consultation and feedback from all stakeholders that are documented, evaluated, and acted upon to ensure closure of the loop;
- A clear system of monitoring with checks and balances; and
- Subjecting its systems and processes to regular internal and external audit and review.

QA Management

To satisfy these premises, WOU's QA is directed from the highest policy bodies such as Council and Senate, and managed by the Deputy Vice Chancellor (Academic) (DVC-A) who chairs the Quality Assurance Committee (QAC) responsible for developing and implementing the University's QA systems and processes. Operationally, a Quality Assurance Unit (QAU), headed by a manager, coordinates and oversees the implementation of QA processes across the University, monitors compliance and recommends continuous improvement measures. The QAU also manages and maintains the quality management system documentation (e.g., QA Policy, QA Manual, Document Procedures, and Records). At the school/departmental level, a Quality Task Force (QTF), headed by the dean, oversees the implementation and review of the QA processes. The QTF reports to the QAC and its continuous improvement plans are regularly reviewed for adoption. All committees operate under clearly defined standing orders with proceedings, and decisions are duly recorded through meeting minutes that are digitally archived.

External peer review and intervention, an important component of WOU's QA system, is firmly integrated into the QA management procedures. At the institutional level, WOU has an international advisory board that provides advice on broad directions and practices that are benchmarked to international norms. It also subjects the entire system to external program accreditation and institutional audit by regulatory authorities. For the development, review, and assessment of curriculum, WOU invites several external experts including:

- Advisory peer groups for the development (and review) of program curricula to ensure academic standard benchmarking and industry relevance;
- External content writers to increase the academic talent pool;
- External course assessors to provide expert assessment on the content of individual courses;
- External examiners to ensure that appropriate exit standards are set for our assessment and exam systems; and
- External program assessors to provide expert review of program structure and relevance.

QA Policy

The University's Quality Assurance Policy (QAP) has the following objectives:

- To develop the QA framework, procedures, and performance indicators in support of the University's vision and mission;
- To inculcate a quality culture and ensure all members of the University community take responsibility for the quality and standard of their work performance;
- To rigorously and continuously monitor to ensure that the policies are implemented fairly and effectively; and
- To develop effective feedback mechanisms that enable the QAC to make informed decisions for improving the quality standards in a timely manner.

To achieve these objectives, WOU has an overarching policy document, the *Standard Operating Procedures (SOP) Framework* that requires all schools/departments to document their respective processes and procedures, based on a prescribed format. This ensures essential information is consistently provided and disseminated to all relevant stakeholders.

As mentioned, the QAU maintains and updates digital records of all the University SOPs. For building a shared responsibility towards a quality culture and ensuring transparency across all levels, digital versions of all SOPs are posted on the staff Intranet portal. Current systems and practices for managing quality can be improved only through the active involvement and ownership of all relevant stakeholders in the University.

EXAMPLES OF QA PRACTICES AT WOU

The following provides a brief description of the way QA is reflected through practice in a number of key academic processes covering program development and approval, course material development, staff training, course presentation, assessment, and the continuous monitoring and review processes.

Program Planning and Development

As seen in Figure 13.2, the development of new programs at WOU involves a double-loop internal approval process before submission to the MQA for approval:

- The Outlined Program Proposal (OPP) benefits from input by members of the external advisory peer group (consisting of senior academics and industry leaders) that ensures academic benchmarking and industry relevance.
- The University Senate and Council critically appraise the Detailed Program Proposal (DPP) before the approved version is used to prepare the MQA submission documents.

An initial evaluation report from the MQA Panel provides the opportunity for the school to react to issues raised. Based on a satisfactory assessment outcome, the MQA approves the program for launch.

Program Accreditation

In the semester prior to the appearance of the first batch of graduates, the program undertakes the MQA full accreditation

Figure 13.2 QA intervention during program planning and development

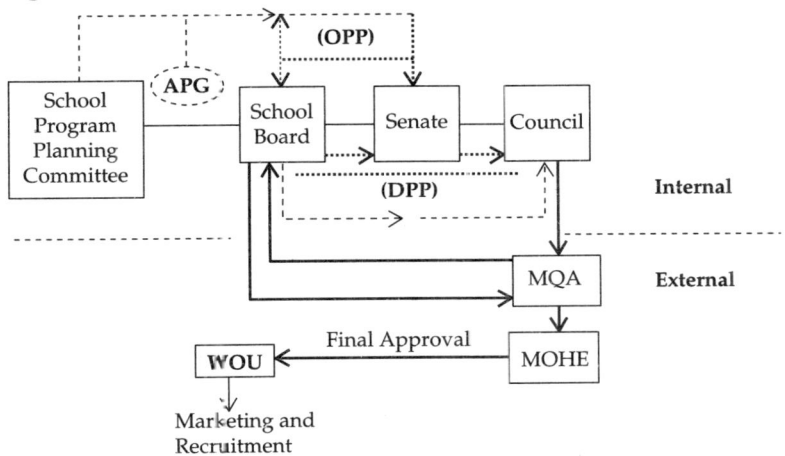

Source: Wawasan Open University (2007a, p. 3).
Note: APG (Advisory Peer Group), DPP (Detailed Program Proposal), MQA (Malaysian Qualifications Agency), MOHE (Ministry of Higher Education), OPP (Outlined Program Proposal).

exercise. At WOU, this is preceded by an evaluation of the program performance by an external program assessor (EPA), a senior academic holding the rank of Professor, as part of its internal QA process. The EPA's independent report is submitted directly to the Vice-Chancellor's office. Submission documents addressing the nine areas in the COPPA guide also include a Self-Review Portfolio. The MQA Panel conducts one or more field visits to audit the QA systems and processes, interviews all stakeholders and reports on their assessment. A satisfactory report leads to the award of Accredite Program status by the MQA and its posting on the MQA Register.

Course Development and Production

As shown in Figure 13.3, course development at WOU uses the Course Team (CT) approach. Each team includes an external writer with subject expertise, one or more internal academics, a course designer and a Web programmer as appropriate. External

Figure 13.3   QA intervention during the course development process

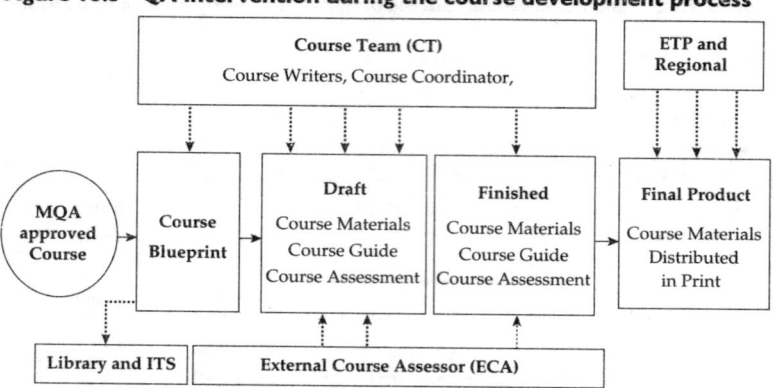

Source: Wawasan Open University, (2007b, p. 4).
Note: ITS (Information Technology Services), ETP (Education Technology and Publishing).

writers are used extensively to increase the subject matter talent pool. A senior academic (Associate Professor and above) in the same field from another University is appointed as the external course assessor (ECA). Based on the MQA approved syllabus, a blueprint is produced following consultation between the course writer and the CT. The blueprint is reviewed by the CT and revised as appropriate before circulation to the librarian (to ensure availability of any text books required), Information Technology Services (to ensure timely installation of required software), and the ECA for critical review.

Once the blueprint has been accepted, the writer proceeds with course material development. The draft of each unit is reviewed by the CT and subsequently forwarded to the ECA for comments. The written comments from the ECA are reviewed by the CT, and where appropriate, guide subsequent refinements. This is repeated for each unit as the development progresses until completion. Multiple interactions between the Course Team and the ECA often occur. All the revised drafts are then submitted to the ECA to enable the preparation of the Final ECA Report. Detailed minutes are taken for all deliberations and decisions made at CT meetings. The final drafts of the unit contents, following endorsement by the ECA, are forwarded to the editorial and graphic design staff for further processing, and the final products are signed off by the coordinator of the CT prior to subsequent production.

At the completion of the development, a Course Development Report incorporating all ECA reports and meeting notes is submitted to the School Board. Once approved, it is sent to the DVC-A, who critically reviews the development for compliance with SOPs and examines the final product before making a recommendation to Senate as to whether the course materials can be used for presentation.

Prior to the start of the semester, the assigned Course Coordinator (CC) must also populate the Course LMS with supplementary materials, quizzes, and other relevant resources.

Learner Support

At WOU, the part-time tutor is a key element of the learner support provision. Figure 13.4 presents WOU's tutor selection, training, and management system. The minimum qualification for a tutor is a relevant Honors degree. Currently, 75 percent of WOU tutors hold Master's or doctoral degrees. New tutors receive training to develop competencies in tutoring, telephone tutoring as well as assignment marking and commenting skills. The performance of all new tutors is monitored by CCs

Figure 13.4 Tutor training and management system

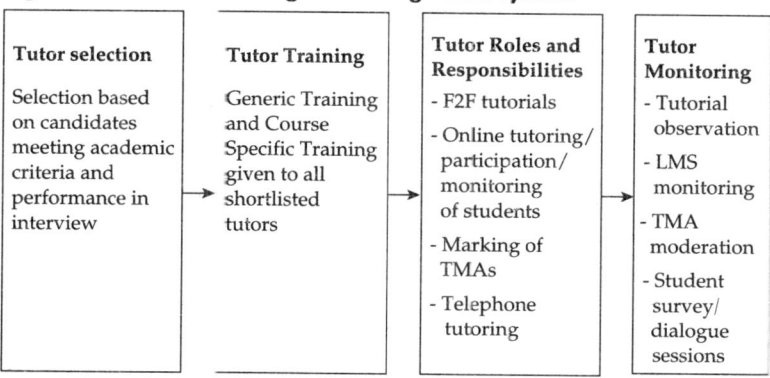

Tutor selection	Tutor Training	Tutor Roles and Responsibilities	Tutor Monitoring
Selection based on candidates meeting academic criteria and performance in interview	Generic Training and Course Specific Training given to all shortlisted tutors	- F2F tutorials - Online tutoring/ participation/ monitoring of students - Marking of TMAs - Telephone tutoring	- Tutorial observation - LMS monitoring - TMA moderation - Student survey/ dialogue sessions

Source: Wawasan Open University (2007c, p. 5).

at the first tutorial and again at the third tutorial. Following the marking of assignments, samples from each tutor are monitored by the CC and feedback is provided with guidance and counseling assistance as appropriate. The performance of the tutors in the LMS is also monitored by the CC, and those who do not provide satisfactory service to the students are flagged and counseled. At the end of each semester, the CC submits an evaluation report for each tutor, and this is used as the basis for reappointment offers.

During the fourth tutorial in each semester, senior academic managers (deans and DVCs) conduct formal dialogue sessions with different groups of tutors, and separately with different groups of students at the regional offices. Issues and feedback raised are documented and submitted to the registrar who then compiles a summary report that is discussed at the following meeting of the Heads of Academic Units (HAUs) chaired by the DVC-A. Decisions are taken at this level, where appropriate, to complete the loop, or the matter is escalated to the Management Board or Senate for further action.

At the same fourth tutorial, a course-specific survey questionnaire is also administered to the students to obtain feedback on the quality of course materials, performance of tutors at tutorials and assignment marking, library, and online support via the LMS. The Education Technology and Publishing (ETP) unit collates all feedback, and reports are made available to the individual Course Coordinators for incorporation into their end of presentation reports as well as the HAU meeting.

Assessment and Examination

The overall system of assessment and examination is outlined in Figure 13.5. As mentioned previously, QA oversight of assignments and examinations involves internal peer vetting followed by the intervention of external examiners, who review the exam paper and the marking guide. The internal examiner and the dean sign off accepted changes suggested by the external examiners prior to printing.

The commercial plagiarism detection software *Turnitin* is available for students to check the originality of their assignments

Figure 13.5   QA in assignment management

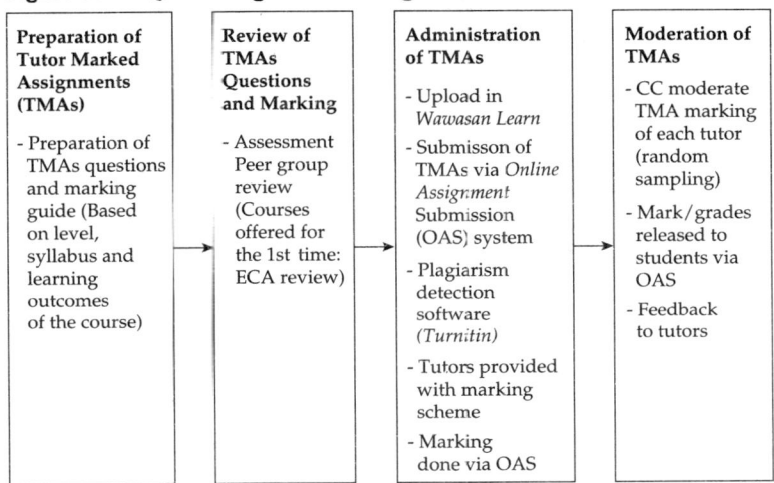

Preparation of Tutor Marked Assignments (TMAs)	Review of TMAs Questions and Marking	Administration of TMAs	Moderation of TMAs
- Preparation of TMAs questions and marking guide (Based on level, syllabus and learning outcomes of the course)	- Assessment Peer group review (Courses offered for the 1st time: ECA review)	- Upload in *Wawasan Learn* - Submisson of TMAs via *Online Assignment* Submission (OAS) system - Plagiarism detection software *(Turnitin)* - Tutors provided with marking scheme - Marking done via OAS	- CC moderate TMA marking of each tutor (random sampling) - Mark/grades released to students via OAS - Feedback to tutors

Source: Wawasan Open University (2007d, p. 3).

before final submission. This promotes academic integrity amongst the University's major stakeholders.

Marking of assignments is handled by the tutors. Samples of the marked assignments are monitored by the CC to ensure marking consistency, reduce bias, and ensure that adequate comments and feedback are given to the students.

Exam papers are produced and printed under tight security. The examination is conducted under proctored conditions. Marking and moderation are carried out according to pre-scribed SOPs. Each marker is assigned approximately one hundred scripts to reduce errors caused by fatigue. From each marker, 5 percent or 12 scripts (whichever is higher) are selected from the low, medium, and high mark range for monitoring by the internal examiner. A maximum total variation of 5 percent for the whole paper is acceptable. Beyond this, remarking is mandatory.

The final processing of the examination results takes place at the Standardization and Award Meeting Chaired by the DVC-A to ensure consistency of decision-making across schools. Results are released only after Senate approval. The

Figure 13.6   QA in the management of examinations

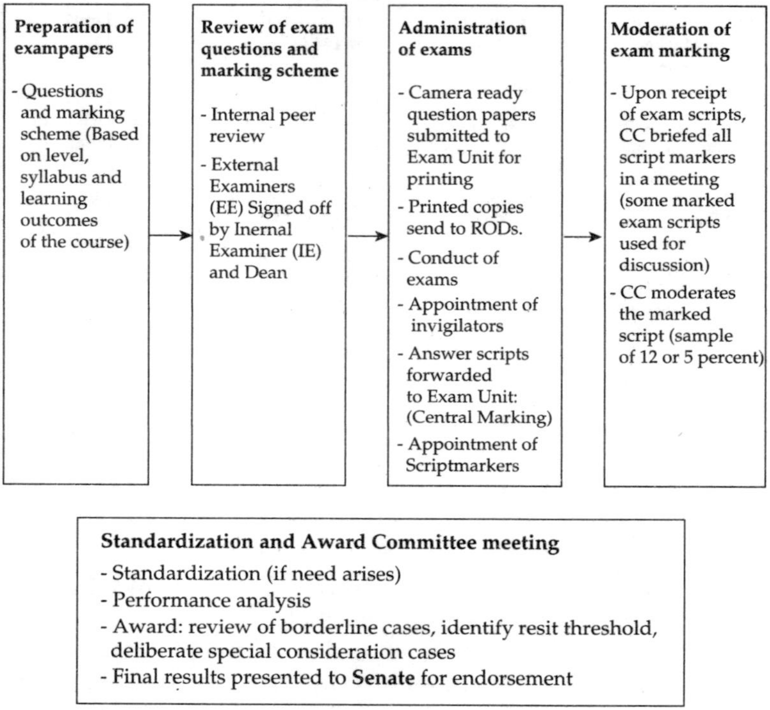

Preparation of exampapers	Review of exam questions and marking scheme	Administration of exams	Moderation of exam marking
- Questions and marking scheme (Based on level, syllabus and learning outcomes of the course)	- Internal peer review - External Examiners (EE) Signed off by Inernal Examiner (IE) and Dean	- Camera ready question papers submitted to Exam Unit for printing - Printed copies send to RODs. - Conduct of exams - Appointment of invigilators - Answer scripts forwarded to Exam Unit: (Central Marking) - Appointment of Scriptmarkers	- Upon receipt of exam scripts, CC briefed all script markers in a meeting (some marked exam scripts used for discussion) - CC moderates the marked script (sample of 12 or 5 percent)

Standardization and Award Committee meeting
- Standardization (if need arises) - Performance analysis - Award: review of borderline cases, identify resit threshold, deliberate special consideration cases - Final results presented to **Senate** for endorsement

Source: Wawasan Open University (2007d, p. 6).

overall system for examination management is shown in Figure 13.6.

Staff

New staff members undergo an orientation program following which they are assigned to mentors who provide advice and assistance as needed. Academic staff members are also required to undertake a formal seven-module ODL Core Competency Training program that leads to internal certification.

Staff performance is assessed through an annual key performance indicator-based appraisal system covering all key functional activities. Annual increment and bonus payments are performance-based, and clear policies on appointment and promotion are laid-down and implemented in a transparent manner.

ISSUES AND THEIR SOLUTIONS

Program Development and Approval

WOU's initial QA protocol for program development, adopted from a self-accrediting university overseas, consumed too long a "product-to-market" timeline. As the WOU Council meets only twice a year, the need for the OPP to be approved by Council before the DPP development could begin caused undue delay. This meant that the internal process took at least seven months. As part of our continuous improvement effort, the WOU Council delegated the OPP approval to Senate, prior to the DPP development. The "product-to-market" time has been reduced by 3–4 months without any significant loss of rigor.

The aims, educational objectives, and program learning outcomes at WOU are all linked to the delivery of education to working adults within a lifelong learning framework. The MQA's External Panels who evaluate WOU programs come with conventional university mindsets. The imposition of norms applicable to conventional universities is often neither relevant nor reflective of quality practice in ODL systems. We engage in continuous dialogue with MQA panels to share the experiences and measures employed by established open universities, including the UK Open University, Hong Kong Open University, and Indira Gandhi National Open University. This has led to a more informed and liberal approach by the MQA when reviewing ODL programs and processes.

Program Delivery

ODL delivery systems rely heavily on the use of information technology (IT). The penetration of IT infrastructure at the students' end is outside the University's control. The availability and costs of software and bandwidth to students has impeded

WOU from using advanced delivery platforms to more effectively deliver its courses. The IT facilities at the main campus and regional offices have been continually upgraded. Most students have benefited from the upgrade and complaints have been significantly reduced.

The introduction of a new delivery medium (CD-ROM) and student support features (OAS, *Turnitin*) suffered from insufficient prior notification and communication to raise early awareness among the students. Learning from these experiences, new SOPs have been put in place and sufficient lead time is now provided through pilot projects before full implementation of new technology.

Insufficient training provided to speakers and tutors using technology to deliver lectures or tutorials caused unhappiness among students during the early years. The assumption that speakers from F2F universities have the necessary lecture delivery skills, especially when dealing with remote audiences, was not valid, and prior training was found to be necessary.

Academic Staff

Staff shortages in the Malaysian education sector, associated with the phenomenal growth of HEIs in the past decade, have led to recruitment difficulties, high staff turnover, and job-hopping. Importing overseas staff, once an option, has become more challenging. Ongoing efforts to recruit academic talent from the local and overseas sources are in place, and newly recruited younger staff are required to undertake comprehensive training designed to develop key ODL competencies. They are also supported to undertake doctoral studies.

The generally higher workload and shortage of research opportunities that is associated with a new primarily teaching institution does not make WOU particularly attractive to younger academics. To create a more attractive work environment for the academics keen on research and publications to drive career advancement, WOU has set up the Institute of Research and Innovation. It is allocated a modest budget to provide research seed-funding for the staff.

External Peers

Although the system of external peer review is in place, WOU's experience over the past four years is that it is individual performance that determines the quality of the intervention as well as any added value. On occasion, the performance of the external peer group has been found inadequate, as the individuals appointed do not share the same quality culture or practice. We now explain our expectations more clearly before appointing external peers. Where it is subsequently found out that expectations are not met, we negotiate mutually acceptable terms of withdrawal.

LESSONS LEARNED

The most important element when developing institutional quality is leadership and the commitment of senior management. Without them "walking the talk," the task becomes extremely challenging. The leaders must be passionate and committed to the development of a quality culture by their example. This commitment must be reflected in the provision of adequate resources to bring about changes in mindset, supported by the establishment of an appropriate QA management unit.

To develop a quality culture in an organization, it is very important for proper training to be provided and made readily available. The required resources are considerable, as it involves changing the way existing staff think and act. Inertia and resistance is inevitable, more so if the staff are already set in their ways. It is important for the agents of change to consult with and benefit from the experience of peers from more established institutions. There must be adequate training for staff at all levels to ensure that they have the necessary competence and skill set to deliver on the quality that is expected of them no matter which specific activity it involves.

There must be proper planning and communication with the relevant stakeholders prior to the introduction of quality enhancing changes to ensure their "buy-in." For example, at WOU the provision of the online submission of assignments was implemented over three semesters. It was piloted using two courses in the first semester, made available on a voluntary basis in the second semester and only became mandatory from the third semester onwards.

CONCLUSION

As it begins its fifth year of operation, Wawasan Open University is still very much a "work-in-progress" in its efforts to become a high-quality institution. Much has been accomplished during the past four years, but more needs to be done. As a private not-for-profit institution, the University faces many challenges in trying to put in place a QA system that ensures that its students enjoy a quality learning experience that over time will become its unique competitive advantage.

Quality comes with a cost, and the need to maintain quality in the face of rising operating costs places severe pressure on the University's other mandate of keeping fees affordable.

In the near future, the University will be progressively moving into an e-Learning environment which will no doubt bring with it additional QA challenges. However, given the lessons learned over the first few years of operation and the emergence of a discernable quality culture at WOU, we can look forward to the future with confidence and excitement.

REFERENCES

Malaysians Qualifications Agency. 2008. *Code of Practice for Programme Accreditation*. Malaysia: MQA. Available online at http://www.mqa. gov.my/en/garispanduan_coppa.cfm and accessed on September 4, 2012.

———. 2009. *Code of Practice for Institution Audit* (2nd edition). Malaysia: MQA. Available online at http://www.mqa.gov.my/en/ garispanduan_copia.cfm and accessed on September 4, 2012.

———. 2011. *Guidelines to Good Practices: Open and Distance Learning*. Malaysia: MQA. Available online at http://www.mqa.gov.my/ garispanduan/GGP%20ODL.pdf and accessed on September 4, 2012.

Wawasan Open University, 2007a. *Programme Planning and Development Process*. Malaysia: WOU.

————. 2007b. *Course Development Process*. Malaysia: WOU.

————. 2007c. *Tutor Management*. Malaysia: WOU.

————. 2007d. *Assessment and Examination*. Malaysia: WOU.

Wong, T.M. 2011. *Wawasan Open University's QA System and Processes in the Context of the QA Framework for Higher Education in Malaysia*. Paper presented at International Workshop on Quality Assurance in Higher Education hosted by Beijing Normal University, 28–29 April. Beijing, China.

Virtual University of Pakistan | 14

Naveed Akhtar Malik

INTRODUCTION

Pakistan is a populous country of nearly one hundred and eighty million people (Ministry of Population Welfare 2011), making it the sixth largest country in the world. It has a relatively young population with a growth rate that shows no signs of slowing down. In 2007, the last year for which accurate statistics are available, the number of people in the 15–30 year age group stood at nearly 40 million (Federal Bureau of Statistics 2011), and it is estimated that the current figure for this age group is well over 50 million. However, the provision of higher education in the country has not kept pace with the rapidly increasing population. Up until 2001/2002, there were less than 60 universities and degree awarding institutions in the country, with the majority being in the public sector. The Higher Education Commission (HEC) was established in 2002 as the successor to the University Grants Commission, and a renewed focus on higher education became visible in the following years. Nine years later, in 2011, the total number of universities and degree awarding institutions stands at one hundred and thirty two, with nearly half (59) in the private sector (HEC Statistical Information Unit 2010). The number of higher education institutions is still well below requirements, and it is estimated that only 5 percent of the college-age group actually finds a place in these institutions. This figure is to be compared with figures of 30 percent and higher in more developed countries.

The HEC is the regulator for higher education in the country, and the principal body through which recurring and development grants are distributed to public sector universities. In its initial years, the HEC concentrated primarily on the upgrading of physical infrastructure in the universities, along with a massive human resource development (HRD) effort that was meant to produce a large number of PhDs who would then become the core faculty of the universities. There was excitement felt across the sector, as pursuing an academic career seemed lucrative for the younger generation, and the universities initiated many new programs.

At this time, it became apparent that the unbridled offering and expansion of academic programs by universities could very quickly become unmanageable unless a quality assurance (QA) framework was established to ensure the validity of these programs. The HEC, therefore, turned its attention toward quality issues in higher education.

Universities in Pakistan are chartered by the government (either provincial or federal), and until recently there was no accreditation of their programs. It was understood that the academic councils of the universities performed this important function, so that there was no need for any external accreditation or QA. A few professional bodies, however, exercised varying degrees of regulation and accreditation in their related programs. These included (Higher Education Commission 2011):

- Pakistan Bar Council (PBC)
- Pakistan Council of Architects and Town Planners (PCATP)
- Pakistan Engineering Council (PEC)
- Pakistan Medical & Dental Council (PMDC)
- Pakistan Nursing Council (PNC)
- Pakistan Pharmacy Council (PCP)
- Pakistan Veterinary Medical Council (PVMC)

The HEC has now developed a Quality Assurance Framework (HEC Quality Assurance Agency 2011) and laid-down important QA criteria, such as those for faculty hiring. As part of this framework, the HEC has started to offer consultations to the universities for developing QA criteria and standards. Quality Enhancement Cells (QECs) are being established at all universities and the HEC is making funding available for this purpose. In the first instance, QECs will ensure that all university programs are properly documented and all support services follow established procedures. QECs will also develop self-assessment manuals and guidelines for use inside the universities. In the final stage, QECs will coordinate with the respective Accreditation Councils that are responsible for external QA.

The Accreditation Councils that have been established by the HEC under its Quality Assurance Framework include (Higher Education Commission 2011):

- National Accreditation Council for Teacher Education (NACTE)
- National Agricultural Education Accreditation Council (NAEAC)
- National Computing Education Accreditation Council (NCEAC)
- National Business Education Accreditation Council (NBEAC)

This effort is in its initial stages, and universities are slowly developing capacity in developing and implementing QA frameworks. All these efforts have concentrated on conventional universities and, so far, no QA/accreditation framework for distance education (DE)/e-learning exists in Pakistan.

VIRTUAL UNIVERSITY OF PAKISTAN

In order to overcome the capacity shortage in higher education, the Virtual University of Pakistan (VUP) was established as a public-sector university in 2002 with a clear mission to provide equitable access to quality higher education to aspiring students across the entire country. It is Pakistan's only university based completely on the innovative use of modern information and communications technologies (ICTs), and uses ICTs as a force-multiplier to overcome the shortage of qualified faculty. Using a judicious mix of broadcast television and the internet, VUP has established itself, in just nine years, as a significant provider of high-quality education. In terms of outreach, VUP has over one hundred and eighty campuses in more than 95 cities that are used to provide infrastructure for studying at the University (computers, broadband connectivity, and lecture viewing facilities). VUP owns and operates 15 of these campuses, while the rest have been established through a vibrant public–private partnership. The Government has approved the setting up of one VUP campus in each district (administrative division) of the country that will add an additional one hundred and forty four campuses over the coming years. With a current enrollment exceeding seventy thousand students, VUP is now one of the largest public-sector universities in the country.

VUP is a modern DE institution that offers formal education using a semester-based system of studies. All instruction is carried out using video lectures developed in the University's own digital recording studios by the best faculty that the nation has to offer The lectures are then either broadcast over the University's own four free-to-air satellite television channels or made available through inexpensive DVDs[1] or freely through the University's OpenCourseware site[2]. Student–teacher interaction is facilitated by a comprehensive learning management system that allows students to interact with their tutors and also to download and submit assignments and quizzes. Examinations are conducted by the University in a formal proctored environment at centers established across the country for this purpose. Students are required to be physically present at these centers to appear for their examinations.

The University offers four-year Bachelor's programs in the areas of Computer Science, Information Technology, Business Administration, Public Administration, Management, Marketing, Business & Information Technology, Accounting and Finance, Commerce, Mass Communication, Psychology, Banking & Finance, and Education. In addition, the University also offers Master's programs in Computer Science, Information Technology, and Business Administration.

This chapter introduces the approach taken by VUP to implement a total quality system, and showcases the assessment system of the University as one specific example of this approach.

---

[1] A three-credit course comprising 45 hours of video lectures can be purchased for about US$0.80 from the University bookshop: http://bookshop.vu.edu.pk
[2] http://ocw.vu.edu.pk

QUALITY ASSURANCE AND ENHANCEMENT MEASURES AT VUP

VUP has established a Quality Enhancement Cell (QEC) as per the directive of the HEC. The QEC reports directly to the rector. The absence of any established standards for DE/e-learning in Pakistan has been discussed with the HEC, and it is obvious that VUP will have to be at the forefront of efforts to formulate and establish these standards. By following a rigorous policy of documentation of its programs from its very inception, VUP is well ahead in self-assessment as compared to much older universities that are only now beginning to undertake QA and quality enhancement (QE) efforts, and is well poised to establish a continuous QA/QE cycle. This section of the chapter examines key QA/QE measures taken by VUP.

QA/QE System Leveraged by Technology

From its inception, VUP leveraged information technology (IT) to the maximum, in terms of both the delivery of education and management. The staff had been inducted from conventional institutions and had very little exposure to DE/e-learning. This was treated as an opportunity, and a continuous capacity-building or training scheme was instituted. The operational systems had to be devised from scratch, since there was no instance that could be used as a role model. In order to determine and establish standards and procedures for the University, it was decided that all departments would use IT as much as possible for their functions, and this use of IT would be systematic. This decision is at the heart of the quality policy at VUP.

All functional requirements of the various University departments are reviewed by a management committee and then submitted to the IT department for induction into the University's management system (henceforth referred to as VIS: Virtual Information System). Once inducted, these tasks can be performed only by strictly following the laid-down IT procedure, and no manual bypass is allowed. The following requirements for QA are thereby ensured:

- All repetitive tasks have to be defined and rules and procedures established (setting standards).
- Tasks are then programmed into VIS.
- Tasks can then be performed only through VIS, thereby ensuring that standards are followed.
- Database log files provide non-repudiable evidence of all actions.

In parallel with VIS, a comprehensive Virtual University Learning Management System[3] (VULMS) has been custom developed that includes all required academic and administrative functions. VULMS allows students to view video lectures, read associated material, download and submit assignments, participate in quizzes and discussions, and interact with their tutors, all within a framework that eliminates the need for e-mail. All academic material that is placed on VULMS has its own defined quality procedures that must be followed before it can be incorporated into the system.

On the administrative side, students perform their course selections, generate and download fee vouchers, and view their grades and accounts at any time, without the need for interaction with any official. Service features under the "Student Services" section of the LMS include:

- Examinations Department
  - o Applying for Rechecking of Papers
  - o Applying for Partial Transcript
  - o Applying for Final Transcript
  - o Applying for Degree
- Registrar's Office
  - o Request a Duplicate ID Card
  - o Request a Semester Freeze
  - o Request a Semester UnfreezeRequest a Migration Certificate (to another University)
  - o Request a Campus Change (within the VUP network)
  - o Request Re-Admission
  - o Perform Course Selection (for a new semester)

---

[3] http://vulms.vu.edu.pk

- Accounts
  - o Apply for a Need-Based Scholarship
- General
  - o View the Student Handbook
  - o Access the HEC Digital Library
  - o View the student's personal Scheme of Studies

For example, when the student performs course selection for any semester, VULMS has complete knowledge of all applicable rules as well as the student's record and thereby ensures all rules and regulations are always followed.

It is important to realize that, by laying down the rules and procedures for these tasks and then programming and inducting them into the University's IT system, the maintenance and application of standards in every aspect of the University's functioning are ultimately ensured.

QA Procedures Specified as Rules

The initial set of statutes, rules, and regulations for VUP has been laid-down in the Virtual University Ordinance (Government of Pakistan 2002), which may be supplemented by the Board of Governors from time to time. Rules are formulated for all activities performed by VUP and provide the basis for standards and procedures to be followed. For each activity, the basic QA lifecycle at VUP is as follows:

- The precise series of steps required to execute the activity are listed.
- Any conditional paths through the series of steps are identified..
- Standard operating procedure (SOP) for the activity is developed, debated, and finalized.
- The SOP is reduced to a set of programmable specifications.
- The activity is programmed into VIS/VULMS and staff is trained.
- After a successful parallel run of at least 1–2 months, manual processing is dropped.
- The process is repeated for further refinements.

The detailed rules and procedures developed for each activity are published, and all items pertinent to students are available in the student handbook (Virtual University of Pakistan 2010). The gradual transition of most of the University's functions to VIS/VULMS has allowed more attention to be paid to individual problems, while all routine matters are handled by the system. This has significant consequent benefits to the University and leads to much higher student satisfaction.

For example, previously, a student request to change from one campus to another (inter-city or intra-city) was a process undertaken by the registrar's office, under the following rules (Virtual University of Pakistan 2010):

- A student may request for campus change up to a maximum of once per month with proper reasons.
- The University requires a minimum of three working days to process the request after completion of the laid-down procedure.

The student would submit the request by e-mail, often to the wrong e-mail address, and then start complaining about the slow response from the office. When the e-mail finally found its way to the concerned official (an Assistant Registrar), he/she would then ask the student for a reasonable justification, unless already provided by the student, which would form part of the official record. The Assistant Registrar would then confirm via e-mail, from the original campus, whether there was any reason why the transfer should not take place (e.g., lecture DVDs that needed to be returned by the student to the campus library). After receiving verification from the campus (which itself could take several e-mail queries and reminders), the Assistant Registrar would then change the status of the student in VIS and also inform the Accounts department. The Accounts department would then update its own records (in VIS) so that subsequent tuition fee share was sent to the correct campus (campuses are not allowed to collect any fee directly from students).

It is obvious that this process was tedious, nontransparent and prone to human errors and omissions. After developing the new QA process, the campus transfer request now operates as follows:

- A student submits a campus change request through his/ her VULMS account.
- VULMS generates a campus change alert for the campus from which the student is transferring, which is sent as a system-generated e-mail and also placed on the VUP portal for the concerned campus.
- If the campus does not offer any objection to the transfer (i.e. does nothing), the transfer is automatically effected in VIS and all concerned systems, including financial system, and updated, immediately after the prescribed three days.

The new system has introduced a high level of efficiency, consistency, and satisfaction into the workings of the registrar's office. The need for human intervention has been eliminated, and it has freed the registrar's office to concentrate on individual cases that cannot be handled systematically.

CASE STUDY: STUDENT ASSESSMENT SYSTEM

In order to establish VUP as a provider of quality education, it was decided that the doors of the University's "virtual" class-rooms would be open to all, thereby showcasing the high-quality of its video lectures. Similarly, it was decided that homework, while being a necessary and important component of the teaching process, would not contribute disproportionately to the final grade; mid- and final-term examinations would be the major components. It became a cornerstone of VUP policy to assess student attainment in a manner that would allow third party evaluation and auditing, and would be acceptable to peer institutions as a valid measure of student performance. All semester examinations would therefore be conducted at properly supervised examination centers where students would be required to appear in person.

In the early years of the University, the student body was small (the initial cohort was five hundred students) and the outreach was manageable (28 campuses in 18 cities). As a result, semester examinations were conducted with pen and paper in a conventional environment. The role of IT was minimal, and consisted basically of getting the question papers to the exam centers on time. Student answer sheets were then returned to the University by courier service and were graded, collated, and entered into the electronic system to be displayed in the students' grade-books. The process was tedious, error prone, and vulnerable to outside factors such as misplacement of answer sheets by the courier. As the number of courses, students, and campuses grew, the problem of conducting examinations nationwide in a consistent and transparent manner within a reasonable amount of time started becoming intractable.

It is instructive to look at the magnitude of the problem. The University currently has an enrollment of over seventy thousand students with more than forty thousand active in any given semester, and the enrollment is increasing with every admission cycle. Each student registers for one to six or seven courses each semester. Since the University admits students during both Fall and Spring semesters, examinations have to be conducted for all courses every semester.

The makeup of the VUP student body mix adds another complicating dimension to the problem. There are a large number of working professionals enrolled who demand that examinations be conducted after-hours or on weekends so that they do not have to take leave from their offices, which may or may not be granted. They are even willing to appear for examinations in more than one course in a day. Full-time students, on the other hand, prefer a gap between their papers, and those who have to commute to their exam center from neighboring towns prefer an exam time closer to mid-day so that they can travel at reasonable times.

Flexibility in the scheme of studies, whereby students can choose to carry more or less than a full course load, is an added complication. No design effort is capable of generating an exam schedule acceptable to all. Much more alarming was the fact that the number of days required to conduct semester examinations was continuously increasing, having serious implications for the academic calendar. The basic problem that needed to be addressed was: "How could the University conduct mid- and final- term examinations in a secure, convenient, time-limited, and efficient manner?", given the complexity of the problem.

In 2007–08, a complete review of the assessment system was undertaken and the possibility of using IT to solve the problem was considered. An analysis of the process proceeded along the following lines:

- Many exam sessions were being conducted with partially full exam halls, since each session catered to a single course.
- To reduce the number of exam days, it was necessary to run exam centers at full capacity.
- This implied that one exam session would have to cater to more than a single course.
- If more than a single course were scheduled during an exam session, the possibility of conflicts increased.

The above analysis led to possible solutions as follows:

- The total number of examination seats required for any exam was known beforehand.
- If students created their own individual examination schedules, there would be no complaints or conflicts.
- This could only be possible if the question papers were different for each and every student.
- Distinct question papers could only be generated by using a question bank approach.
- The question bank would have to be large enough and designed in such a way that valid question papers could be generated.
- Examinations would have to be conducted using computers to eliminate all problems associated with moving paper parcels around.
- A new way of grading student attempts would have to be found, otherwise brain-fatigue could easily overwhelm a grader who was presented with hundreds of different examination papers.

The academic heart of the problem was the question bank that would have to be created for every subject. While the IT Department got busy with the software specifications and development, the academic staff was subjected to intensive training on question development. The following process was developed for submitting a question to the bank for any particular course:

- Staff would be required to develop and submit 3–5 multiple-choice, short or long essay questions every week of the semester for their respective courses.

- Senior staff would review and either accept, accept with modifications, or reject the submissions.
- The Head of Department or a senior faculty member would review the questions and approve them for induction into the question bank.

Each question had to be classified by lecture, topic, taxonomy, difficulty level, etc. Certain simplifications had to be made to the process to reduce the dimensionality of the problem. Thus, the difficulty level, the time taken to solve the problem and the number of marks allocated to the problem were initially considered synonymous. Screenshots for the submission of essay-type and multiple-choice type questions are shown in Figures 14.1 and 14.2 respectively.

From a conduct perspective, the sequence of events followed by the VUP e-assessment system is outlined below. All steps indicated are carried out through the IT-based system and no

Figure 14.1   Essay-type question definition

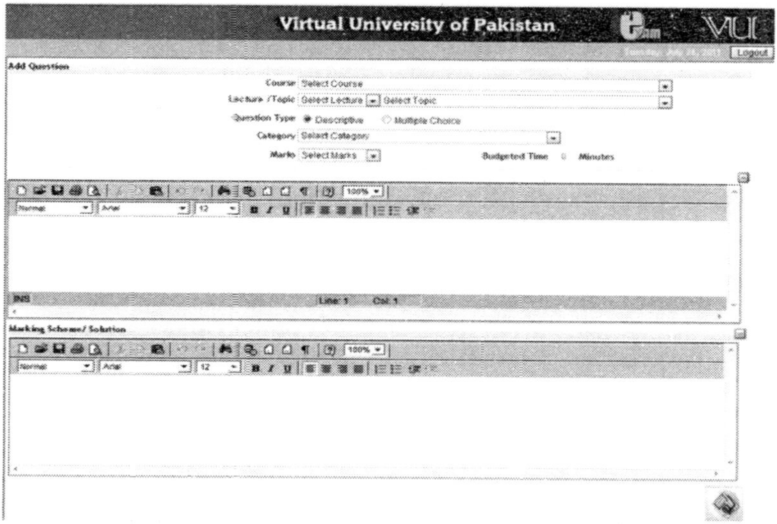

Source: Screenshots from the Virtual University Examination System (http://exams.vu.edu.pk) which is accessible only through the University's internal Network).

Note: Grading Rubric at bottom.

Figure 14.2   Multiple-choice type question definition

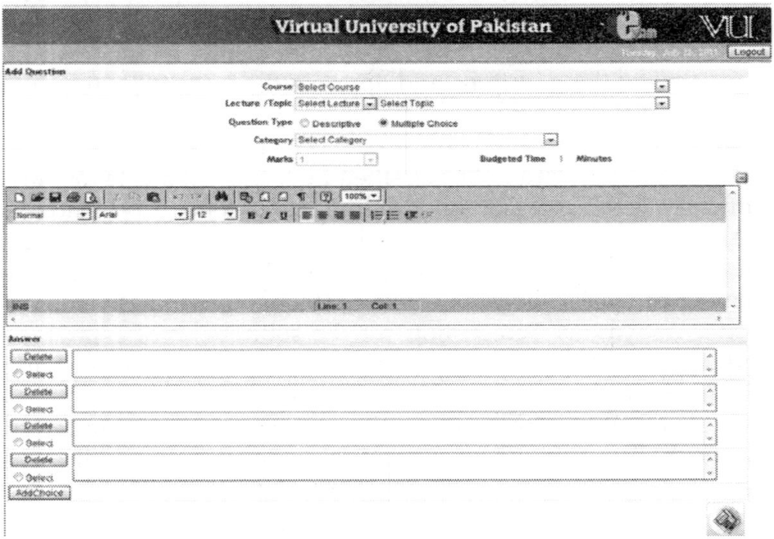

Source: Screenshots from the Virtual University Examination System (http://exams.vu.edu.pk) which is accessible only through the University's internal Network.

Note: Possibility to adjust number of choices at bottom left.

manual intervention is necessary, other than the provision of required parameters.

In the planning phase, the Examinations department:

- Determines the total number of exam seats required (number of students x number of subjects);
- Updates its databank about exam center capacity; and
- Determines and defines a grid of exam centers, number of exam days (including weekends), duration of the exam, number of sessions in a day, and session timings.

Manual input is required only to specify the starting date of the exam and the session timings. Examination dates are then announced and students invited to create their exam schedules (also known as datesheets) on a first come, first served basis. This announcement is placed on the VULMS Notice Board, sent via e-mail and also via SMS to all students. On receipt of this information:

- Students log in to http://datesheet.vu.edu.pk using their VULMS credentials;
- Select the city where they wish to appear;
- Choose a designated exam center within the city from the list provided by the system;
- Their current semester subjects are displayed;
- For each subject, they select the day/date and session time and confirm; and
- Finally, they are able to obtain a printout of their examination entrance slip (Figure 14.3) that carries the name, student ID, city, center, photograph, examination schedule, and a list of instruction that they need to follow.

Once the students have created their datesheets, the system knows the exact question paper requirements for each session/day/center.

The next step is the generation of question papers for which the academic staff provides parameters on a course-by-course basis. These include such items as the number of lectures to

Figure 14.3 Exam entrance slip after creating datesheet

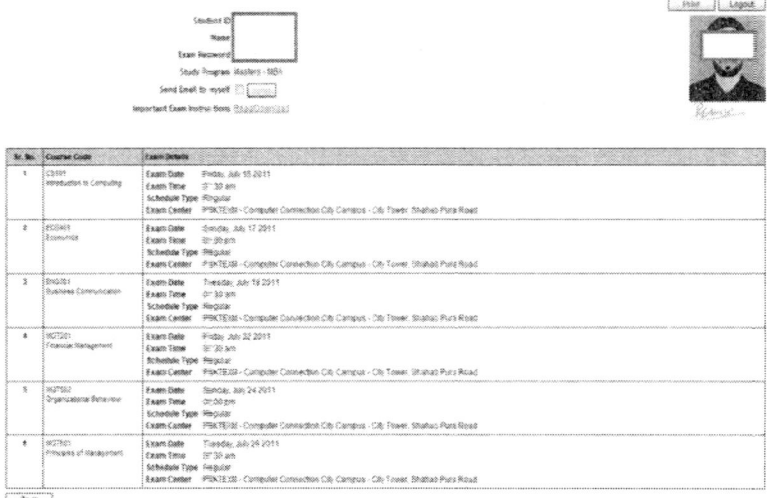

Source: Screenshots from the Virtual University Examination System (http://exams.vu.edu.pk) which is accessible only through the University's internal Network.

Figure 14.4   Setting question paper criteria

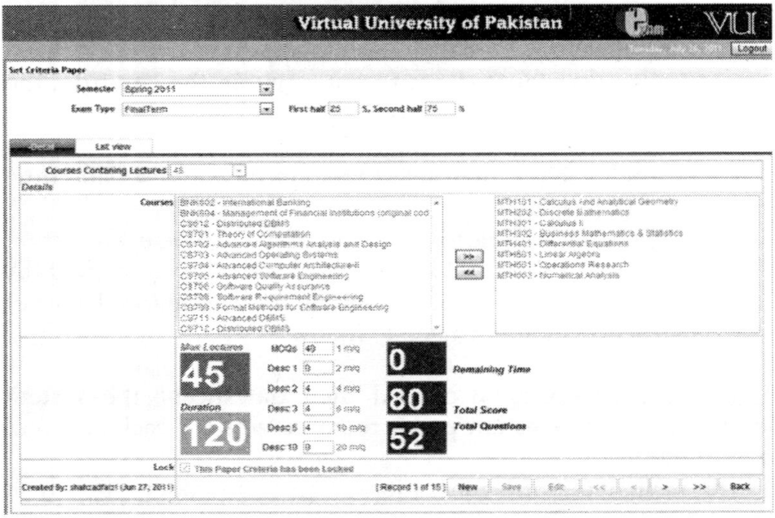

Source: Screenshots from the Virtual University Examination System (http://exams.vu.edu.pk) which is accessible only through the University's internal Network.

be covered in the examination (for mid-term examinations; final examinations always cover the entire course), the duration of the test, the number of multiple-choice questions (MCQs) to be used in the paper and the number of essay type questions to be included (Figure 14.4). The system then generates the required number of question papers from the respective subject's question banks.

When the question papers have been generated the following process takes place:

- All e-papers for an exam center are collected into e-bundles, which are doubly encrypted;
- The e-bundles are dispatched by email to the exam centers;
- Examination software and encrypted data are loaded onto the exam center servers;
- Decryption passwords are sent to invigilators by e-mail/ SMS 5–10 minutes before the session;
- Data for a single session only is unlocked;

- Students log in and receive their own distinct question papers according to their self-designed datesheets;
- Tamper-procf and encrypted files containing students' answers are sent to VUP by e-mail on a daily basis;
- On receipt, MCQs are graded by the system;
- Essay-type attempts by students are collected and sorted by question;
- All attempts for one question are given to a single tutor for grading alorg with a grading rubric (Figure 14.5);
- When grading is completed, the results are collated and declared; and
- Students receive results in their VULMS grade-books.

Other than the actual grading of the essay-type questions, all other steps in the process are fully automated and require no manual intervention other than deciding when to declare results.

A careful analysis of the system reveals the benefits obtained by adopting this approach. The noise about unreasonable exam schedules has completely disappeared, since students now create their own datesheets. Examinations are conducted in an efficient

Figure 14.5 Grading an essay type question with a rubric provided by the question bank

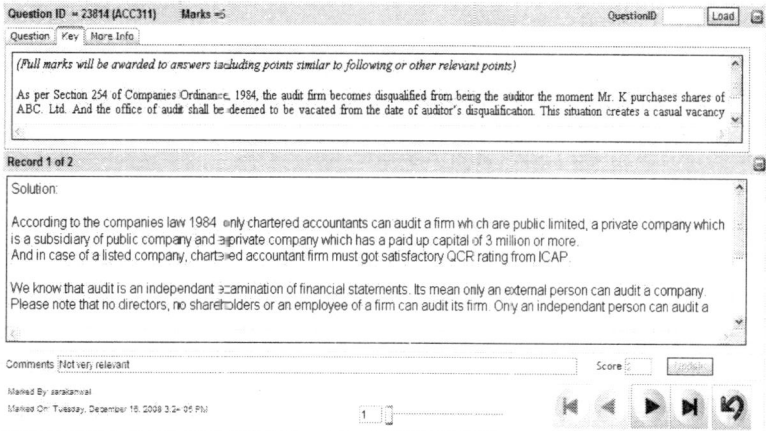

Source: Screenshots from the Virtual University Examination System (http://exams.vu.edu.pk) which is accessible only through the University's internal Network.

Note: Absence of any student identification.

manner with exam centers being run at near-full capacity, and the number of days required has been drastically reduced. The use of unfair means has essentially been eliminated, since students receive distinct question papers (different questions, not just permuted questions). Exam data is encrypted at all times (even during the grading process; the human readable form is limited to the computer screen) and is completely tamper-proof.

The VUP examination system was presented at the LINC 2010 conference (Malik 2010) and generated a considerable amount of interest, including suggestions for patenting the system. The consistency of examination quality established by this system is being appreciated nationwide and many traditional and distance universities including the Allama Iqbal Open University have requested presentations about the system with a view to adapting it for their own use. The major impediment that other universities need to overcome is the generation of their own question banks, since VUP courses are, in general, more rigorous and intensive, and the VUP question bank may not be directly suitable for them.

The main quality thrust is now focused on the question bank itself. All questions in the bank carry a "use-counter" that indicates how many times it has been used in any examination. After the use-counter exceeds a defined threshold, the question is marked as retired. In the meantime, fresh questions are added to the bank on a continuous basis as described earlier, and the question bank is kept refreshed. The quality of the questions is kept under constant review, and senior academics from outside VUP perform this critical function.

LESSONS LEARNED

The establishment of internal QA standards, where none exist at national level, is a painstaking process that requires a consistent Systems Development Life Cycle approach. In universities, especially in DE institutions, many disparate departments need to work together as a cohesive whole to ensure that the organization functions efficiently. This has implications for the setting of standards. Although a single department may use international best practices as the starting point for its operating procedures, this must not be finalized in isolation. A collaborative inter-departmental approach produces the best results. This is especially true when IT is brought into the picture. While well-designed and implemented IT-based systems allow a near-perfect implementation of standards and procedures, the risk of mass-level failures also exists due to unforeseen side effects, especially if systems are designed in isolation.

The buy-in from top management for the induction of QA in any organization is critical for its success. Leadership at the top level also has to act as a change agent and, unless it converts the rank and file of the organization into adopting a system-centric quality approach, the risk of failure is high. In a real-time system, it is imperative that all tasks be performed as and when required and the very natural human tendency for "batch" processing must be avoided at all costs.

CONCLUSION

A careful study of the VUP's approach to assuring and enhancing quality in its academic and service offerings indicates that the University relies on a system analysis and design approach for all of its procedures. By implementing these procedures in the form of an ever-growing IT-based system, the University has been able to ensure quality in every aspect of its work.

The task was made easier by the fact that the University is young and started from a greenfield state. There was no legacy baggage and IT systems were developed, albeit in primitive form, from day one. All staff, in all departments, had to internalize the IT-based way of working and the system has matured into a well-oiled machine. As a matter of policy, all procedures that can be automated are automated, and where human intervention can be eliminated, it is. The primary objective is to establish a student-friendly environment where students are the determinants of their own path through the University's academic system. If the near elimination in the number of complaints received from students is any indicator, the approach is working very well. In the absence of established standards for distance education/ e-learning in Pakistan, the VUP is well poised to help define and establish these standards.

REFERENCES

Federal Bureau of Statistics. 2011. Federal Bureau of Statistics. Available online at http://www.statpak.gov.pk/fbs/sites/default/files/population_satistics/publications/pds2007/tables/t01.pdf and accessed on September 4, 2012.

Government of Pakistan. 2002. "Virtual University Ordinance," in The Gazette of Pakistan: 1150–1176. Islamabad: Government of Pakistan.

HEC Quality Assurance Agency. 2011. Quality Assurance Framework. Available online at http://www.hec.gov.pk/InsideHEC/Divisions/QALI/QualityAssurance/QADivision/Documents/Quality%20Assurance%20Framework.pdf and accessed on September 4, 2012.

HEC Statistical Information Unit. 2010. Number of Universities and Degree Awarding Institutions in Pakistan (1947-2009-10). Available online at http://www.hec.gov.pk/InsideHEC/Divisions/QALI/Others/Pages/UniversitiesDAIs.aspxand accessed on September 4, 2012.

Higher Education Commission. 2011. Quality Assurance Agency. Available online at http://www.hec.gov.pk/InsideHEC/Divisions/QALI/QualityAssurance/QualityAssuranceAgency/Pages/AccreditationCouncils.aspx and accessed on September 4, 2012.

Malik, N.A. 2010. "Assessment of Large Student Cohorts: In a Formal Distributed Learning Environment," Paper presented at the MIT LINC 2010 Conference, 23rd May, 2010, Cambridge, USA. Available online at http://linc.mit.edu/linc2010/proceedings/plenary-Malik.pdf and accessed on September 4, 2012.

Ministry of Population Welfare. 2011. Population Dynamics. Available online at http://www.mopw.gov.pk and accessed on September 4, 2012.

Virtual University of Pakistan. 2010. Student Handbook. Available online at http://handbook.vu.edu.pk and accessed on September 4, 2012.

PART 5

Outcomes and Performance Measurement

Open University of Sri Lanka[1]   15

Uma Coomaraswamy

INTRODUCTION

The University Grants Commission (UGC), together with Committee of Vice-Chancellors and Directors (CVCD) of Sri Lanka's university system, established a formal quality assurance (QA) system in 2002. Frameworks for institutional and subject reviews were developed and published as the *Quality Assurance Handbook for Sri Lankan Universities* (UGC-CVCD 2002). The *Academic Procedures Handbook* (UGC-CVCD 2004a) is a compilation of Codes of Practice, providing a reference point for all universities and covering all main aspects of academic standards and quality of education. Subject Benchmark Statements (UGC-CVCD 2003) and the *Sri Lankan Credit and Qualification Framework* (UGC-CVCD 2004b) were also developed as complements to the *Quality Assurance Handbook*. They serve as useful building blocks in support of quality and academic standards. The *Credit and Qualification Framework* indicates how a particular university's award/ qualification and the level and volume of credits relate to a national qualification and credit "standard," whilst the coverage and content of a particular program of study leading to that qualification can be matched with the relevant subject benchmark statement. Subject Benchmark Statements describe the nature of the subject area, the expected attributes and capabilities of award holders, and minimum standards for the award of the degree.

Activities that have been initiated by CVCD/UGC are at present coordinated by the Quality Assurance and Accreditation Council of the UGC (QAAC-UGC) as an interim measure. The

[1] Assistance given by Dr Gayathri Jayatilleke, secretary, QA unit of OUSL, is gratefully appreciated.

Standing Committee on Quality Assurance and Accreditation serves as an advisory body to the QAAC-UGC, which conducts institutional and subject reviews in all universities. In the first cycle, during the period 2004–2010, all 15 universities underwent institutional reviews, and 95 percent of Subject Reviews were completed. A second cycle will commence by the end of 2011. The QAAC-UGC also conducts awareness workshops for faculty staff in all universities and "Reviewers" training programs to create a national pool of academics and professionals to carry out the uphill task of Reviews in all universities. The QAAC-UGC is a member of the International Network for Quality Assurance Agencies in Higher Education (INQAAHE) and the Asia Pacific Quality Network (APQN).

In compliance with the covenants in the Project Administration Memorandum (ADB 2003), the Distance Education Modernization Project (DEMP 2006–2009) developed a QA framework and a national evaluation process for QA in distance higher education, Accreditation Standards, Performance Indicators, and Evaluation Criteria for Accreditation of Distance Higher Education Institutions and Programs. Based on a concept paper (Coomaraswamy 2009) outlining the need for establishment of national accreditation agency for Distance Higher Education, including its structure and composition, strategic goals and mechanism of accreditation, the Minister of Higher Education (MoHE) has approved for consideration by the Cabinet the setting up of an accreditation agency for Distance Higher Education.

A Cabinet paper on the establishment of the "Sri Lanka Qualification Framework, Quality Assurance and Accreditation Board" which will have under it the "Quality Assurance and Accreditation Agency" and the "Qualifications Framework Agency" is now being considered by the government. This will also incorporate QA in distance higher education.

OPEN UNIVERSITY OF SRI LANKA

The Open University of Sri Lanka (OUSL) was established in 1980 as a single mode distance teaching university (MoHE 1980). It is one of the 15 in Sri Lanka's university system under the purview of the UGC, with the same legal and academic status as any of the other universities. OUSL's central campus is situated in Nawala, a suburb of Colombo, and its outreach centers are located throughout the island to provide geographical access to all who aspire to achieve higher education. It differs from the conventional universities in that it is fee levying, with its own admission policy and pedagogy in teaching and learning. It receives all capital funds and two-thirds of its recurrent funds from the government.

OUSL's organizational structure is similar to that of the conventional universities. The principal executive officer is the Vice-Chancellor. The University is managed by Council, Senate, and Faculty Boards. It has four faculties, nineteen departments and more than ten administrative divisions.

The University has over two hundred academic staff, more than 50 academic support staff, more than 50 administrative staff, more than 70 technical staff and over four hundred nonacademic staff, based mainly at the central campus with a few at the regional centers. It also recruits more than six hundred visiting staff, appointed to deliver academic programs in the various centers of the regional network.

The learner profile is diverse, distributed across the country, and represents all major ethnic communities. Any person over 18 years of age can enroll in a program of study. The male to female ratio is 53:47. Approximately 80 percent of students claim to be employed in varied occupations. With its thirty thousand students enrolled, compared with around seventy-six thousand students in the conventional universities, OUSL has nearly 30 percent of the tertiary enrolment in national universities.

OUSL offers a wide range of academic, professional, vocational, and in-service training programs through over 40 programs and more than nine hundred courses, leading to Foundation, Certificate, Diploma, Bachelor's Degree, Post-graduate Diploma, Master's and Doctoral Degree. A readily accessible and progressive ladder of opportunity is provided through its flexible lateral entry/exit points to enable students to progress at their own pace. Previous qualifications and work experience are recognized, and appropriate exemptions are offered.

The Distance Education Modernization Project (DEMP), supported by the Government and funded by the Asian Development Bank (ADB 2003) from 2003 to 2009, along with government efforts to develop infrastructure and provide high-end technology and other facilities to promote distance education (DE) at a national scale, have helped to substantially modernize the OUSL. This allowed the OUSL to:

- Improve its infrastructure facilities in the central campus and in all the regions;
- Provide short-term and long-term training of academic, administrative, and technical staff as an integral part of capacity enhancement;
- Provide instructional, design, and material development for online programs/courses;
- Enhance institutional research capacity;
- Enable digitalization of the library; and
- Provide ICT infrastructure facilities to establish and support the OUSL Management Information System.

DEMP also facilitated the development of the national QA framework for higher education through distance learning. Overall, DEMP has made a major contribution toward helping OUSL to move to online learning.

This chapter focuses on OUSL's quality management structure and QA practices pertaining to learning and teaching processes. It reveals how such quality practices have evolved into a national framework for Distance Higher Education, leading to the internationalization of Sri Lankan standards and performance indicators, in collaboration with Commonwealth of Learning (COL) and UNESCO. The challenges faced in this arduous journey and the positive initiatives taken, along with lessons learnt, are discussed.

QUALITY ASSURANCE AND ENHANCEMENT MECHANISM AT OUSL

Quality Management Structure

The governance structures and processes adopted by OUSL have much in common with conventional universities. However, OUSL has incorporated various modifications to reflect the processes associated with an open and distance learning (ODL) system in ensuring academic quality.

The University has developed an integrated framework to ensure quality in all its operations, and implements a strategy for continuous quality enhancement. Relevant stakeholder groups were represented in the policy formulation process, to encourage ownership and to facilitate feedback and openness. Monitoring procedures ensure that the policies and plans of the university are implemented, evaluated, and improved. The strategies, policies, and procedures have a formal status and are publicly available (OUSL 2000; 2006c; 2011).

At the apex, Council is responsible and accountable for overall administrative and quality management of all operations of the university. The Senate, an exclusive academic authority, is responsible for QA with regard to curricula, bylaws, and rules and regulations pertaining to examinations and awards. Faculty Boards are responsible for academic administration of the departments and quality in all academic operations.

OUSL adopts a committee structure whereby different committees, standing committees and ad hoc committees of the Council, Senate, and Faculty Boards meet regularly to monitor and evaluate academic and administrative operations of the University so as to ensure "service quality." Staff representation at different levels in governance structures provides opportunities for staff for formal involvement in institutional matters and creates a sense of belonging among all members.

Quality products, processes, and outcomes are achieved via the vision, mission, objectives, strategies, and short- and long-term action plans in the five-yearly corporate plan, along with commitment to internal mechanisms for regular monitoring and feedback at all levels (the Council, Senate, Faculty Boards). They also provide a direction and a path for the working staff, as the QA mechanisms developed by OUSL are integrated into institutional processes. The QA Unit of the University is tasked with coordinating all quality-related activities and liaises with QAAC-UGC. Initiatives are currently underway to strengthen the QA Unit with additional roles and responsibilities.

OUSL's commitment to quality management reflects its willingness to voluntarily subject itself to external reviews as part of the QAAC-UGC programs for 2003–2010. This comprises an institutional review and two program reviews, by DEMP (2006–2010) and the Commonwealth of Learning Review and Implementation Model (COL-RIM). In 2010, the recommendations of the review panels were used to draw up action plans to strengthen the weaknesses identified and undertake a number of initiatives. Action plans developed from the recommendations of COL-RIM are currently being discussed by the Senior Management Committee.

Quality Interventions in Distance Learning and Teaching

There are clear processes and procedures for program and course design and development. The setting and maintenance of academic quality and standards, and legal aspects with respect to the program and courses, are provided through university academic rules, regulations, bylaws, and manuals of procedure.

The University assures the quality of its programs, courses, and awards through a linked system of processes which ensures that:

- A program is relevant, matches established benchmarks and standards and is approved by the academic authority;
- The design of individual courses which make up a program contribute to achieving the overall program objectives;
- Courses are designed and developed according to the standard formats, with periodic reviews and improvement;
- A robust system of assessment is in place; and
- Relevant learner support services are in place.

Program Design

Programs are developed based on a need analysis exercise, including market research and liaison with industry and professional bodies.

- Specific committees of the Senate ensure that the proposal meets appropriate academic standards and is feasible in technical and resource terms.
- The main strategies used to review programs are recommendations of the review panels of external institutional review and subject reviews by QAAC-UGC; formal and informal feedback from key external stakeholders, internal staff, and students.

Course Design and Development

Instructional design and the course development process are governed by the Senate, which approves technical and production standards in course design and development.

- The use of subject benchmark statements ensures curricular content of a subject is appropriate to the level and nature of the award.
- The Course Team concept is adopted, which ensures that all stakeholders are involved in the development and production of course materials from the beginning of the course development cycle.
- QA criteria for instructional design include appropriateness of learning and teaching strategies, appropriateness of content, and utilization of effective media for delivery.
- Monitoring the progress of course development occurs at both departmental and faculty level. Draft course material is pilot-tested with potential learners, peers, and tutors using questionnaires relating to the relevance, accuracy, credit rating, comprehensiveness of content covered, learner-friendliness, and interactivity. Feedback is used in the improvement of course materials.
- Formal mechanisms are in place for approval, evaluation, monitoring, and periodic review of programs and courses.

Assessment

Assessment is integral to teaching and learning strategies.

- Different forms of assessment strategies (formative and summative) appropriate to achieving the stated learning outcomes of the individual program/course are decided at the course development stage and stated in the course material.
- Details regarding examination procedures and processes are laid-down in the Senate and Council approved Manual of Procedures for Continuous Assessments and Final Examinations. In addition, there are bylaws relating to examinations and degree awards, including for ensuring confidentiality and integrity in setting, marking, and record keeping.

- Assessment standards are benchmarked against the practices in the national university system.
- Completion, retention, and progression rates are monitored on a continuous basis. The University provides periodic reports on student performance and completion, published in the annual reports submitted to the Parliament.

Learner Support

Learners are supported through the provision of a range of tutoring functions at a distance to facilitate their holistic progression.

- Self-learning study material, a network of libraries, a network of regional and study centers in 28 locations, computer facilities, audio-visual aids, practical sessions, day schools/ discussion classes, tutorials/workshops/tutor clinics are provided for learner support.
- Other means of learner support include pre-admission counseling, orientation programs for new students, a system of personal counseling, provision of a "Learning to Learn" course, advice through the career guidance unit, health care, bursaries and scholarships for needy students, peer support through approved societies, and temporary residential facilities.
- There are formal and informal mechanisms to obtain feedback, both qualitative and quantitative, from students and tutors on learner support. This includes course evaluations and surveys on counseling and the contact sessions.

DEVELOPMENT OF NATIONAL QA FRAMEWORK AND PERFORMANCE INDICATORS

QA Framework for ODL

OUSL developed a QA framework for ODL in collaboration with COL and the UGC, involving members of the OUSL Governing Council, Senate, Faculty Boards and departmental staff, Quality Assurance Standing Committee of the UGC and international resource persons from the COL. The development involved customization of the generic guidelines in the QA Framework for Distance Education Institutions, published by the Asian Association of Open Universities (AAOU and COL 2002). The QA Framework identified ten criteria:

1. Policy and planning;
2. Human resources provision and development;
3. Management and administration;
4. The learners;
5. Program design and development;
6. Course design and development;
7. Learner support;
8. Learner assessment;
9. Technology for learning; and
10. Research, consultancy, and extension services (OUSL 2006a).

To facilitate the preparation of self-evaluation reports for reviews, a *Manual of Self Evaluation for ODL Institutions* (OUSL 2006b) was also developed.

The impact of DEMP triggered an educational paradigm shift, whereby some conventional universities transformed into dual-mode institutions. At the same time, DE providers increased several fold (Coomaraswamy 2007). The rapid expansion of higher education, together with the exponential growth of DE providers, posed a threat to the quality of distance higher education. The overriding concern was to assure the quality of DE programs and establish parity between DE and the conventional educational delivery system.

DEMP Administration Memorandum (ADB 2003) mandated the development of a national QA framework, finalization of a policy including guidelines and evaluation criteria for accreditation of DE programs, and the development of a National Accreditation Policy and Plan for distance higher education. In response, extensive discussions were held involving OUSL academics and administrators, along with representatives from conventional universities offering external degrees, other public and private higher education institutions using DE and professional training institutions. OUSL played a lead role in devising a national QA framework, which spans the full range of distance delivery methods, from print-based to online learning that is sufficiently flexible to accommodate particular and individual institutional solutions in a contextually specific environment.

Performance Indicators

To move beyond external QA processes toward engaging institutions in guided self-assessment that promotes the development of a reflective culture of quality, performance indicators (PIs) were developed for each "Accreditation Standard." New and emerging ODL systems, in their formative stages, could use the indicators to develop appropriate corrective actions (Coomaraswamy and Abeywardena 2007; Coomaraswamy, Kondapalli, and Abeywardena 2008).

Realizing the potential use of PIs for reflective self-evaluation, COL decided to partner with DEMP to internationalize the indicators developed in Sri Lanka. The indicators were reviewed and revised to be more generic at a Pre-PCF5 workshop in July 2008 by a team of distance education and quality assurance experts from Canada, Hong Kong, India, Jamaica, Malaysia, Pakistan, Papua New Guinea, South Africa, Sri Lanka, United Kingdom, and UNESCO. The resulting PIs, a collection of case studies of best practices and a glossary are contained in the *Quality Assurance Toolkit for Distance Higher Education Institutions and Programs,* published as a trilateral partnership between Sri Lanka, COL, and UNESCO (Kondapalli, Hope, and Coomaraswamy 2009).

Experience in the Use of the "QA Toolkit"

The QA Toolkit is used by OUSL for self-evaluation. Many new DE providers in Sri Lanka are using it as a check-list to guide them in developing and offering their programs. QAAC-UGC is adapting relevant standards and indicators for use in institutional and subject reviews in the conventional university system, in large measure because of its objectivity in evaluation.

An international roundtable was organized by COL and OUSL in August 2010 to obtain feedback on the use of the PIs. It was attended by representatives from APQN, INQAAHE, AAOU, OUSL, Open University Malaysia, Institute of Agro Technology and Rural Sciences, Jamaica, Nigeria, and Consultants from COL. The main purpose of the roundtable was to share the experiences of the users of the QA Toolkit and invite suggestions from QA agencies for further improvement.

OUSL has used the standards for self-review of the Bachelor of Science Degree conducted by the Faculty of Science, where all five departments are involved (Jayasooriya 2010), and the Master of Arts in Teacher Education (MATE), conducted by a department in the Faculty of Education (Karunanayaka 2010). The strengths and concerns/limitations of the "QA Toolkit" indicated by OUSL are given in Table 15.1.

Table 15.1   OUSL's experience with the QA toolkit

Strengths	Concerns/Limitations
MATE	
• Very comprehensive • Cover all aspects of program • Clear guidelines given • Scoring system indicated • Activity of self-review very challenging and demanding • Sense of satisfaction and motivation • Self-evaluation through critical reflections was an interesting exercise • Identification of areas to be improved	• Overlapping of certain standards • Repetitive nature of certain standards • Difficulty in interpreting certain standards
B.Sc.	
• Standards together with indicators will serve a useful check-list in setting goals • Features like PIs will make self-evaluation and collecting information a lot easier • Having standards in the form of statements makes the process of assessment easier for the reviewers • The standards were able to assess the faculty's commitment to learners, academic standards of the program, curricula, program design and course design • Using the toolkit, a faculty or a department can independently learn about their strengths and weaknesses for improvement	• Some standards are lengthy and a few are repeated • Some standards not sufficiently focused to assess the particular program • Some standards not clear to reviewers who do not have much experience in ODL • Standards did not focus on physical resource requirements • Standards did not focus on obtaining views of employers

Source: Jayasooriya (2010) and Karunanayaka (2010).

ISSUES AND PROBLEMS FACED AND SOLUTIONS

Overcoming Stigma of "Second Rate" Higher Education

OUSL is one of the 15 national universities, with the same legal and academic status as any of the conventional universities. All permanent appointments, confirmations, promotions of staff conform to national policies and are made in accordance with the schemes of recruitment prescribed by the Universities Act and implemented by the UGC. However, during early stages OUSL suffered from the stigma of "second rate" education, because of the unique combination of an "open education policy" and "distance mode" delivery. A major challenge was to ensure parity of esteem with counterpart conventional universities, and through that build confidence in the minds of employers.

Measures Adopted by OUSL Included

- Drawing senior academics and professionals from the conventional universities and professional organizations as external examiners, moderators, authors and editors for course books, tutors for delivery of programs/courses, members of selection panels for recruitment and promotions, project supervisors, and as senior consultants during their sabbatical leave;
- Voluntarily subjecting to external reviews of both the university and departments by the QAAC-UGC;
- Being reviewed by external reviewers from conventional universities and professional bodies, and ensuring that OUSL academics are on review panels for conventional universities; and
- Inviting representatives of QA Units from conventional universities, the QA Standing Committee of the UGC and the QAAC-UGC, together with COL and international experts, in the development of the QA Framework, accreditation standards and PIs for distance higher education.

These measures were of great value in building confidence in the minds of conventional universities, employers, and the public.

Contributing to the Development of a National QA Framework for ODL

The impacts of DEMP were manyfold, resulting in over 25 institutions other than OUSL offering distance, online, or blended learning programs. Many of them were conventional educational/training institutions new to distance teaching. To safeguard the sustainability and credibility of these new and emerging ODL systems, appropriate QA protocols at the national level had to be devised. Further, it was a covenant in the ADB Project Memorandum (ADB 2003).

Three Important Challenges were Faced in Devising a System for Them

- A national QA framework should span the full range of delivery methods (from print to online).
- The framework should include a credible and universally applicable set of QA guidelines, which are at the same time sufficiently flexible to accommodate individual institutional solutions.
- It should be a dynamic QA system that has the capacity to adapt to a changing environment.

The QA Framework for ODL prepared by OUSL (OUSL 2006a) was used as the starting point to develop a national level framework, at a series of workshops drawing representatives from each of the institutions, along with universities and professional bodies. During discussions, DEMP was cautious not to impose ideas on the working groups to avoid any counterproductive effect. It allowed consensus to evolve from the group, and that strengthened the feeling of ownership of the process. In-depth groundwork done by DEMP was enhanced by an appropriate combination of focus group workshops and national level seminars. Transparency and clarity in the approach instilled confidence among the participants.

Capacity-Building

For ODL operations to be successful, it is essential that a Course Team has members with the knowledge and skills to design and develop instructional materials, writing skills, the tenacity to work toward deadlines, and the commitment and dedication to complete the task effectively. This is a serious challenge, as many OUSL academics are recruited from the conventional universities and lack these skills. OUSL has addressed this issue through in-house training and retraining in its Staff Development Center, upgraded by DEMP. It also plans back-up processes to ensure the deadlines for course material production and timely distribution to learners.

Inculcating a Quality Culture

Another significant problem lies in convincing all staff members that quality is a matter of continuous improvement leading toward excellence, and that all institutional members should work together in achieving that excellence. Induction/orientation sessions for new staff include discussions on quality aspects of ODL operations and the importance of all individuals in the institution committing and contributing to a quality culture.

LESSONS LEARNED

The initiative taken by OUSL in first developing a QA tool for the institution, then developing a national QA framework and, finally, internationalizing it across the Commonwealth has presented a number of difficulties. These included financial resources, and the requirement for committed human resources with ODL background, along with motivated contributors, with and without ODL experience, who were willing to share long periods of their time with us. The leadership and commitment of top management and DEMP consultants, combined with pro-active and innovative OUSL academics and administrators, as well as the contributions of other persons representing universities and other tertiary education providers, made this unique exercise a reality. COL, the Sri Lankan Government and ADB-DEMP provided generous financial support. The commitment, participation, and financial assistance of the government, along with international involvement in QA initiatives, are important factors in sustaining quality.

Other useful lessons include the following:

- A QA framework should be "home grown" through a bottom-up participatory process that ensures ownership by those who will be using it.
- To provide direction for the workforce, institutional leadership and commitment is imperative.
- Collaborative, particularly international, ventures enable costs to be shared and add value to the system.

CONCLUSION

Appropriate QA policies and processes, along with complementary governance and management, have ensured that the programs of study at OUSL are academically rigorous and that quality standards are maintained in learning and teaching processes.

Having developed a QA framework for ODL, the OUSL was called upon by DEMP to take a lead role in developing the QA framework for distance higher education at the national level. This led to the development of PIs for institutions, particularly at their formative stage of DE development, to gauge their own performance trends and foster continuous quality improvement in the spirit of a culture of quality. Several useful lessons learned in this evolutionary process have been briefly described.

The QA procedures and practices outlined above took more than a decade to evolve, and are continuously being subjected to review and improvement. Meticulous planning, adequate resources (both financial and human), and commitment from all staff are vital in producing high-quality of products and services, so necessary in achieving excellence.

REFERENCES

AAOU-COL. 2002. *Quality Assurance Framework for Distance Education Institutions.* Bangkok and Vancouver: UNSECO-PROAP and Commonwealth of Learning.

ADB. 2003. Project Administration Memorandum for the Loan to the Democratic Socialist Republic of Sri Lanka for the Distance Education Modernization Project (DEMP). Loan 1999—SRI (SF).

ADB. 2007. "Quality Assurance in Open and Distance Learning—Sri Lankan initiatives," Paper presented at the National Workshop on Performance Indicators for Quality Assurance in Distance Higher Education, Colombo, Sri Lanka, August 14–17.

Coomaraswamy, U. 2009. *National Assessment and Accreditation Council for Distance Higher Education.* DEMP Concept Paper, Colombo, Sri Lanka: Ministry of Higher Education.

Coomaraswamy, U. and Abeywardena, N.S. 2007. "Transforming Higher Education through National Distance Education System: Ensuring Quality," Paper presented at the 21st AAOU Annual Conference, Kuala Lumpur, Malaysia, October 29–31.

Coomaraswamy, U., Kondapalli, R., and Abeywardena, N.S. 2008. "In Pursuit of Excellence in Distance Higher Education: Harmonizing Quality Assurance Systems," Paper presented at the 22nd AAOU Annual Conference, Tianjin, China, October 9–14.

Jayasooriya, T. 2010. "Experiences in using Standards of Accreditation for Internal Quality Assurance proves of Bachelor of Science Degree (BSc) Programme, OUSL," Paper presented at the International Roundtable Conference in Quality Assurance, Colombo, Sri Lanka, August 8–9.

Karunanayaka S. 2010. "Experiences in using Standards of Accreditation for Internal Quality Assurance proves of Master of Arts in Teacher Education (MATE), OUSL," Paper presented at the International Roundtable Conference in Quality Assurance, Colombo, Sri Lanka, August 8–9.

Kondapalli, R., Hope, A., and Coomaraswamy, U. 2009. *Quality Assurance Toolkit for Distance Higher Education Institutions and Programs.* Sri Lanka: DEMP, Vancouver: Commonwealth of Learning, and Bangkok: UNESCO.

MoHE. 1980. *An Introduction to the Open University of Sri Lanka.* Colombo,Sri Lanka: Ministry of Higher Education.

OUSL. 2000. *The Corporate Plan 2000–2005.* Sri Lanka: The Open University of Sri Lanka.

———— 2006a. *Quality Assurance for Open and Distance Learning: Framework for Quality Assurance–Volume 1.* Sri Lanka: The Open University of Sri Lanka, Vancouver: Commonwealth of Learning, and Sri Lanka: University Grants Commission.

————. 2006b. *Quality Assurance for Open and Distance Learning in Sri Lanka: Manual for Self-Evaluation of Open and Distance Learning Institutions–Volume 2.* Sri Lanka: The Open University of Sri Lanka, Vancouver: Commonwealth of Learning, and Sri Lanka: University Grants Commission.

————. 2006c. *The Corporate Plan 2006–2010.* Sri Lanka: The Open University of Sri Lanka.

OUSL. 2011. *The Corporate Plan 2011–2016*. Sri Lanka: The Open University of Sri Lanka.

UGC-CVCD. 2002. *Quality Assurance Handbook for Sri Lankan Universities*. Sri Lanka: Committee of Vice Chancellors and Directors, University Grants Commission and The Open University Press.

———. 2003. *Subject Benchmark Statements*. Sri Lanka: Committee of Vice Chancellors and Directors and University Grants Commission.

———. 2004a. *Academic Procedures Handbook*. Sri Lanka: Committee of Vice Chancellors and Directors and University Grants Commission.

———. 2004b. *Sri Lankan Credit and Qualifications Framework*. Sri Lanka: Committee of Vice Chancellors and Directors and University Grants Commission.

Open University Malaysia[1] 16

Anuwar Ali and Mansor Bin Fadzil

INTRODUCTION

As Malaysia moves into a high-income economy, the need to develop the country's human capital has become the mainstay of the government's policies. This is manifested by the more than three hundred and thirty million US dollars allocated under the economic stimulus package to maintain the country's competitiveness and enhance efficiency of the labor force, as announced by the government in January 2009.

Higher education in stitutions (HEIs), in particular, those involved in open and distance learning (ODL), are well positioned to participate in this agenda, as they cater largely to the working adult population. With a number of ODL providers offering their programs, including those from abroad, there is a genuine concern about the quality of educational provision. As a result, the role played by the Malaysian Qualifications Agency (MQA) in providing systems for ensuring the quality of tertiary education has become critical.

In Malaysia, the national quality assurance (QA) or accreditation framework for higher education is under the purview of the MQA, a statutory body established under the Malaysian Qualifications Agency Act 2007 to accredit academic programs provided by HEIs and facilitate the recognition and articulation of qualifications. It is an independent, autonomous body with a mandate to assess the quality of all tertiary level institutions, including universities. The MQA, formerly known as the *Lembaga Akreditasi Negara* (LAN), was established under the National Accreditation Board Act 1996 prior to its merger with the Quality Assurance Division, Ministry of Higher Education, in December 2005.

---

[1] The authors acknowledge the contributions of Kamariah Mohd Noor, Tina Lim, and Rohaya Ali in the writing of this book chapter.

The MQA administers the Malaysian Qualifications Framework (MQF). The MQF is an instrument that classifies qualifications based on a set of approved criteria and benchmarked against international best practices. It clarifies the earned academic levels, learning outcomes of study areas, and the credit system, based on students' academic load. These criteria are accepted and used for all qualifications awarded by recognized HEIs. Hence, the MQF integrates with and links all national qualifications. It also provides educational pathways through which it links qualifications systematically. These pathways enable an individual to progress, through credit transfers and accreditation of prior experiential learning, in the context of lifelong learning.

A set of guidelines, standards, and codes of practice has been developed by the MQA to help HEIs enhance their academic programs and increase institutional effectiveness through self-reviews and internal assessments, as well as external audits. The codes of practice to which Malaysian HEIs are currently expected to adhere are:

- The Code of Practice for Program Accreditation (COPPA), which governs the practices of institutions in curriculum design and delivery. Program accreditation is based on compliance to this code of practice; and
- The Code of Practice for Institutional Audit (COPIA), which provides a QA system for institutions to conduct self-reviews or for external audit. It contains both benchmarked and enhanced standards for institutional compliance.

The quality evaluation process for both codes covers nine areas, namely:

1. Vision, mission, and learning outcomes;
2. Curriculum design and delivery;
3. Student selection and support services;
4. Assessment of learners;
5. Academic staff;
6. Educational resources;
7. Program monitoring and review;
8. Leadership, governance, and administration; and
9. Continual quality improvement.

The MQA's approach to QA involves program accreditation (two levels: provisional and full accreditation) followed by institutional audit (two main components: self-assessment review and MQA institutional audit) and self-accrediting certification. Information on accredited programs and qualifications awarded by the HEIs in Malaysia can be accessed through the Malaysian Qualifications Register (MQR)[2]. The MQR is the reference point for all accredited programs.

OPEN UNIVERSITY MALAYSIA

Established on 10th August 2000, Open University Malaysia (OUM) is the country's first ODL HEI. OUM is owned by the Multimedia Technology Enhancement Operations (METEOR) Sdn. Bhd., a consortium of 11 Malaysian public universities. Although incorporated as a private university under the Private Higher Educational Institutions Act 1996, the University leverages on the quality, prestige, and capabilities of its strategic partners— the 11 public universities. OUM adopts the motto "University for All," which is consistent with its philosophy to democratize education. This philosophy underlies the belief that education should be made available to all, regardless of time, place, age, and socio-economic background.

OUM began operations in May 2001, with the first intake of learners in August that year. Presently, the cumulative enrolment is one hundred and five thousand, with a total of 57 academic programs offered, ranging from diploma to postgraduate levels. Altogether there are 53 learning centers, of which 48 are located in Malaysia whilst five are overseas, in Yemen, Bahrain, Maldives, Ghana, and Sri Lanka.

As with other academic institutions, the Senate is the highest body responsible for academic matters. The University Academic Management Committee (UAMC) facilitates the implementation of academic policies as endorsed by the Senate, and the Group Management Committee (GMC) oversees the day-to-day running of the University.

---

[2] http://www.mqa.gov.my/mqr/index.htm

In 2006, the University became the first institution to obtain approval from the Ministry of Education, Malaysia, to implement an open entry system. The open or flexible entry system provides an alternative path for prospective learners to gain entry into OUM academic programs through less stringent requirements as compared to traditional requirements set by conventional universities. It takes into consideration applicants' prior learning and work experiences.

The ODL programs conducted at OUM are specifically targeted at working adults who need to look after their families as well as manage their careers while pursuing their studies. OUM thus adopts a blended approach in the delivery of teaching and learning. The blended approach comprises:

- Self-managed learning, where learners study at their own pace using specially developed learning modules (both print-based and web-based);
- Face-to-face interaction, where learners interact with tutors/facilitators at Learning Centers during weekends; and
- Online learning, where learners interact online with peers and e-tutors.

As much as OUM attempts to ensure that the education provided remains affordable to a wide section of society, and that it lends sufficient flexibility to learning, it is firmly committed to and unequivocal in ensuring that the programs offered are of the highest quality. In this regard, as with all public and private institutions of higher learning in Malaysia, OUM ensures that all academic programs and qualifications comply with the MQF.

This chapter will discuss OUM's internal quality management measures, and analyze, through two case studies, how the University has addressed QA issues. It will also offer lessons from these cases that could be useful to other ODL institutions.

QUALITY MANAGEMENT AT OUM

At OUM, the Quality, Research, and Innovation Council spearheads all QA and quality enhancement initiatives. It is chaired by the President/Vice-Chancellor of the University, and comprises the GMC members. The Institute of Quality, Research, and Innovation (IQRI) acts as secretariat to this council.

The University's quality objectives, policy, and the prescribed guidelines and procedures are documented in OUM's Quality Manual[3] that is made available online to all staff. The quality objectives include:

- Obtaining accreditation from the MQA for all academic programs;
- Obtaining recognition from the Public Services Department for all accredited programs;
- Ensuring that the processes and procedures of SIRIM (Standards and Industrial Research Institute of Malaysia) certified departments (Registry, Center for Instructional Design and Technology (CIDT), Center for Student Management (CSM) and Digital Library) meet the MS ISO 9001:2008 standards;
- Inculcating a quality culture among staff through continuous quality improvement efforts, timely dissemination of information, and capacity-building activities; and
- Inculcating the University's shared values amongst all staff.

Based on the Quality Manual, service operation manuals are developed at the departmental level, both for academic departments as well as support centers and learning centers. Each of the individual department's service operation manuals details the structure, processes, standard operating procedures (SOPs), and work instructions for effective and efficient quality management toward achieving the University's quality objectives.

Efforts are consciously and diligently taken to set in place and facilitate the implementation of total quality management university-wide. This pertains to continuous improvement in the provision of quality services and products to the customers, which at the same time, meets stakeholders' expectations and requirements. In this regard, the Quality Unit within the IQRI plays a pivotal role in creating quality awareness and instilling

---

[3] http://iqri.oum.edu.my/oum/index.php?c=iqri&v=art_view&domid=33&
parent_id=388&cat_id=421&art_id=1043&lang=eng&PHPSESSID=0ea4d25b90
bef34b02a46c44498ab8f6

a quality culture among staff, and also in ensuring that both internal and external QA processes are in place via the planning, implementation, monitoring, and review of quality initiatives. It also provides consultancy services to departments/faculties within the University with regard to the development and review of SOPs of all core processes.

Overall, the internal QA system of the University complements and facilitates external QA processes, while the external QA standards and indicators are used to drive the transformation of core internal processes and structures, and provide institutional focus on areas for continuous assessment and quality improvement. QA processes outlined by the national QA body, the MQA, are employed by the University to improve and monitor OUM's performance in all areas related to the provision of higher education to learners. In addition, OUM, being a customer-focused organization, realizes the value and relevance of having ISO certification for core processes central to the University's outcomes, namely the learners' successful attainment of learning outcomes of the academic programs registered for as well as their satisfaction with the learning experience. Consequently, OUM opted for four learner support service centers, i.e., the CSM, the CIDT, Registry, and OUM's Digital Library, to be measured against the requirements of the Malaysian Standards MS ISO 9001:2008. The core processes of these service centers have successfully attained certification.

Recognizing that teamwork and collaboration within and across departments are the mainstay of any quality management system, the University gives due emphasis to holding QA meetings that are attended by quality management representatives from all departments. This facilitates communication and encourages collective and coordinated quality initiatives. Further, internal audits are periodically scheduled, both for departments certified under the MS ISO 9001:2008 standards, as well as those that are not. In order to facilitate internal audits, selected staff across all departments (both academic and support services) are sent for training on the conduct of audits. In-house workshops are also organized to develop human capacity in this area. Following all internal and external audits, findings and recommendations are reported at various levels of management, including the UAMC. Management reviews are held to ensure that the quality management system remains current and meets the University's requirements, the needs of its customers as well as the expectations of all stakeholders. Additionally, the University's quality policy is regularly communicated, both overtly and covertly, to all staff through various channels such as bimonthly staff assemblies, monthly University newsletters, seminar series, as well as management and departmental meetings.

As part of the University's internal QA measures, OUM utilizes the Balanced Scorecard (BSC), a strategic planning and management tool that helps organizations align business activities with their vision and strategy, as well as monitor organizational performance against strategic goals. With emphases on quality, costs, productivity, innovation, and operational efficiency, the BSC focuses on four major components or perspectives:

1. Financial (the financial measures of success of the organization);
2. Customer (customer satisfaction levels that lead to the attainment of financial objectives);
3. Internal process (business processes in place to satisfy the customers); and
4. Learning and growth (the level of expertise of employees that sustains the ability to change and innovate).

In this respect, measures that relate to those four components are developed and reviewed yearly at OUM. Key performance indicators are identified and tied to the budget approved by top management before the beginning of each financial year. The performance of various departments is then tracked against the planned budget quarterly. The yearly management report for the whole University is collated and submitted to the Board of Directors for endorsement. Staff performance with respect to the target is assessed formally on an annual basis. Having used the BSC for about five years, the University has found it to be very useful in clarifying and translating the University's vision, mission, and strategy, communicating and linking strategic objectives and measures, planning, setting targets, and aligning strategic initiatives, and enhancing strategic feedback and learning.

Additionally, since 2009, OUM has periodically conducted self-reviews using the Commonwealth of Learning (COL) Performance Indicators for Distance Higher Education Institutions.

The ten criteria examined are: vision, mission and planning; management, organizational culture, and leadership; learners; human resource development; program design and development; course design and development; learner support and progression; learner assessment and evaluation; learning infrastructure and resources; and research, consultancy and extension services.

The quality indicators proposed for the ten criteria mentioned above help the University to gauge its own performance with respect to inputs, processes, and outputs. While this is not a direct indication of quality or the lack of it, the University finds the indicators beneficial, particularly in striving to meet stakeholders' expectations and requirements as well as to gain their confidence.

TWO CASES: HOW INTERNAL QA MEASURES HELP ADDRESS DELIVERY ISSUES

Even as OUM strives to provide quality education through proper planning and development of its quality management system, it does face shortcomings in the implementation phase. This is where internal QA procedures such as internal audits and feedback from learners enable the University to detect problems at an early stage and allow corrective action to be identified and implemented.

The case studies that follow pertain to two key aspects of OUM's operations, i.e., the supply of learning materials or modules to learners, and a revision in the format for continuous assessment. The background to the quality issues as well as the steps taken to overcome them through self-reviews are presented.

Case I: Supply of Learning Modules

While OUM adopts a blended approach in the delivery of instruction, and all three components of face-to-face tutorials, self-managed learning using modules and online discussion using myVLE (OUM's "home-grown" learning management system) are equally important, learners tend to rely on the printed modules. This is due to the fact that the printed modules are readily accessible at all times as compared to virtual sources of information that require internet access, and face-to-face interaction, which is scheduled for only eight hours per semester for undergraduates and ten hours per semester for postgraduates.

Every semester, all printed learning modules are expected to reach learning centers throughout the country and distributed to all learners before the new semester begins. Timeliness in the delivery of the learning modules is critical to ensure that learners are given sufficient time for self-study prior to the first face-to-face tutorial. This requires accurate and timely information relating to course registration for the upcoming semester.

The Issue

Based on complaints from learners, as well as feedback obtained from learning center administrators, in 2009 there appeared to be a problem in the supply of modules. In some learning centers, there was an under-supply of certain modules and in others there was an over-supply. The under-supply of modules caused some learners to start the semester without modules, while a surplus of modules at certain learning centers resulted in requests for additional office space, previously allocated for teaching and learning, to be converted into module storage areas.

A probe into the issue of under- and over-supply led to the discovery that the number of modules (for some selected titles) received by some learning centers was less than the number of registered learners, due to inaccurate data on the registered number of learners that serve as a reference to print the modules. Shortfalls also occurred during the packaging and distribution stages.

The Self-review

Arising from the issue, a multi-departmental study team comprising key personnel from the relevant departments and led by the Director of the IQRI was established with the following terms of reference:

- To review the current process relating to the supply and demand for modules;

- To identify and analyze the issues and possible root cause(s) affecting the under- and over-supply of modules; and
- To put forward appropriate recommendations to the top management.

The study team examined the module delivery process flow, and identified issues that could be contributing factors to under- and over-supply and possible root cause(s) at various points in the delivery process. Then, based on the data obtained, the team determined actual root cause(s) and recommended solution(s) to the problem/issue.

Based on the reported under- and over-supply of modules, affecting 28 percent of the titles, a study was conducted at selected OUM Learning Centers, i.e. in Kota Kinabalu, Batu Pahat and Kota Bharu. Findings from the study indicated that inaccuracy in the course registration data was due to:

- Last minute "pushing" of several courses;
- Credit transfer requests being entertained even after the registration deadline. This caused the courses registered earlier to be dropped;
- Some courses being excluded in the earlier stage;
- The workload of staff in the faculty (in particular, that of the Faculty of Education and Languages where the number of courses handled are significantly higher than those of other faculties) contributing to oversights in the identification of courses for a particular semester;
- Without giving prior notice, learners choosing to collect their modules at centers different from the ones they had registered in; and
- Learning centers requesting additional back-up copies; where the number far exceeds the recorded number of registered learners at the cut-off date.

Based on the above findings, the following main recommendations were put forth by the study team to the UAMC:

- To develop an effective centralized module inventory system integrated with the i-Campus Student Management System;

- To make the i-Campus system accessible to all relevant departments to view pertinent information;
- To consider printing modules in two batches, according to the registration cut-off dates;
- To review allocation of manpower at the Faculty of Education and Languages; and
- To adopt the MS ISO 9001:2008 approach as a process monitoring tool.

Additional recommendations were put forward with regards to the development of supplementary materials to the learning modules in the form of e-resources such as e-content, Learning Objects, i-Radio, i-Cast, i-Tutorial, and Audio/Video Archives. It was also recommended that myVLE should provide options for progressive downloads, and that offline facilities/devices such as CDs could be made available to learners. It was noted that learners need to be given at least one semester prior notification should policy changes be introduced to improve this process.

During the UAMC meeting on April 7, 2009, the following initiatives were mooted:

- For senior learners, re-registration is to commence one semester ahead, with i-Campus automatically proposing a set of courses for learners to register based on their program of study, after taking into account courses already taken or credits already accumulated;
- Learners to be given a specified timeframe to select the proposed courses online and make payment by a specified date to confirm their registration for the upcoming semester. The data generated from this process to be used to determine the number and title of modules to be printed according to the number of learners and their geographical locations.

Post review initiatives: A Centralized Module Inventory Management System was developed in stages, with the first component being pilot-tested in September 2010 at the Shah Alam Learning Center. Starting from the September 2010 Semester, the distribution of modules is now performed in two batches, taking into account learners who register within the stipulated timeline and those who do not comply. In January 2011, a late registration fee (RM100) was introduced as a deterrent for learners who do not observe the timeframe. In addition, arising from the pilot run of the integrated Centralized Module Inventory Management System and the i-Campus Student Management System, several shortcomings were detected, such as the use of live data from i-Campus that slows down the process of extracting the registered courses. Further, the use of printed forms to scan learner IDs will be revisited, with the ultimate aim of introducing a completely paperless operation.

- Capture information of modules: title, ISBN number, price, etc;
- Enable users at the CIDT (the center responsible for module development and delivery) to update the quantity of modules printed and distributed to the centers, whilst the learning centers will capture the remaining balance of the modules after distributing to learners during registration;
- Enable the matching of the number and list of modules collected with the number and list of registered learners by course;
- Use of barcodes and scanners to minimize human error in data entry and thus speed up the module collection process;
- Generation of reports such as "List of modules available by learning center," "Number of modules for re-printing," etc; and
- Capture and keep track of the stock of modules at all the key points of the process, i.e., from the demand for modules in i-Campus to the warehouse in Kuala Lumpur, to the learning centers and at the CIDT.

Case 2: Assignment Practices

Assignments are an integral part of OUM's assessment system, as they are a component of continuous assessment. Typically, most courses have an assignment component weighting that constitutes 30 percent of the total marks. Assignment questions are given to learners at the beginning of the semester.

The Issue: There are three main issues in the implementation of the assignment component at OUM. The first issue pertains to the quality of assignment questions, the second relates to the timeliness in the feedback given to learners as part of their formative assessment and lastly, incidents of plagiarism in learners' assignments.

First, in regard to the quality of the assignment questions, the University relies on subject-matter experts (SMEs) to formulate a collection of assignment questions, together with a detailed marking scheme to guide assessors in grading assignments. In the past, the emphasis was to look for the "right answers" to the questions posed. However, following a random audit conducted by the UAMC on assignment questions, several shortcomings to this previous practice were identified:

- The questions were not able to elicit answers that could discriminate between learners of varying learning capacities;
- The marks obtained for assignments challenged the logic of continuous assessment, in which high scores in assignments do not necessarily lead to correspondingly high marks in the final examination;
- Because the questions were mostly from the lower levels of the cognitive domain, this assessment mode had limited capacity in moving learners to higher levels of learning;
- There appeared to be cases of plagiarized work among learners from different learning centers. This was not noticed by the assessors, as marking of assignments was performed by face-to-face tutors at the different learning centers. With the widespread use of the Internet, learners were able to "share" their work with their peers throughout the country; and
- The distributed marking of assignments also often resulted in delays in the announcement of the results to learners; thus, learners were not able to get timely feedback on their performance.

The Self-review: From May 2010, the following revised practices were recommended to address the issues:

- Close-ended questions with specific answers were replaced with the use of single open-ended questions in the form of essays, cases, laboratory reports, field reports, reviews, etc., which require learners to present articulated and coherent arguments on a certain issue or problem;

- Incorporation of questions that include higher-order levels of thinking (application, analysis, synthesis, and evaluation) and require answers that are supported by readings and/ or research that promotes information literacy. This enables learners to develop information literacy and critical thinking skills;

- In line with the above, the Institute for Teaching and Learning Advancement (ITLA) was given the responsibility of working closely with faculty representatives to develop rubrics for marking assignments, and also to develop guidelines for the formulation of assignment questions by the SMEs and for answering assignment questions by learners. The guidelines are linked to the topic related to assignments in the OUMH1103 "Learning Skills for Open and Distance Learners" course and uploaded on myVLE and the ITLA webpage; and

- Development of an assignment submission platform to assist learners in the timely preparation and submission of assignments.

Post review initiatives: The Online Assignment Submission Platform was successfully developed in 2010 by METEOR Technology and Consultancy Sdn. Bhd., a subsidiary of METEOR. The system that enables learners to submit their assignments online is incorporated into myVLE, and has been implemented since the May 2010 semester. The Online Assignment Submission platform allows learners to view, download, or print their assignment questions from myVLE. This feature provides ease of access and allows learners to work on their assignments irrespective of their locations or time zones. The online submission feature is another important aspect of the platform. It allows learners to submit their completed assignments in soft copy. It provides a faster way of submitting assignments and enables the learners to save money in terms of printing, binding, and delivering assignments to the learning centers. Upon successful submission, learners can print an acknowledgement slip as proof of submission. Each slip has a unique identification number. Moreover, assignment rubrics are made available in myVLE in order to assist learners during the preparation of assignments. Rubrics comprise a set of criteria and weightings used for assessment and for grading. Assignment templates are also made available in myVLE, and can be downloaded as reference by learners and tutors/facilitators. Finally, a list of FAQs related to online submission of assignments is available in myVLE.

In order to ensure that all assignments submitted are not copied from other learners or plagiarized, the platform has a plagiarism detector feature called the "Similarity Checker" to check for dishonest practices. The checker will determine the percentage of similarities between two assignments and produce a report, which in turn will highlight the need for a review of marks.

The online grading system allows appointed graders to view assignments and mark them according to rubrics or answer schemes prepared by the faculties. Marks entered by graders are automatically saved into the system for further processing before being released. The system also helps provide centralized marking with immediate feedback to promote learning.

LESSONS LEARNED

Several observations can be drawn from the case studies featured in this chapter, which may serve as useful lessons for other ODL institutions. They are as follows:

- Complying with various standards and criteria set by external bodies/organizations is definitely helpful to an HEI in assuring the quality of its products and services. However, HEIs must not rely solely on external mechanisms but need to have in place internal QA/QE processes that enable it to localize needs and priorities. This is particularly so for ODL institutions in whose country QA standards have been developed with the requirements of conventional universities in mind.
- Regular monitoring and review of procedures allow for detection of issues/problems. This facilitates the identification and implementation of both reactive and proactive measures that can close the gap between the quality objectives of the organization and the current state of affairs.

- ODL institutions that rely on technology in the delivery of instruction, and face issues related to technology, can leverage the very same technology to alleviate problems that may surface. It is both effective and efficient to utilize technological tools to overcome technology-related problems.

CONCLUSION AND FUTURE CHALLENGES

QA and quality enhancement is the mainstay of all efforts in meeting customer and stakeholders' expectations and requirements. Understanding one's own strengths and weaknesses as a flexible education provider helps tremendously when it comes to challenging the efficiency and effectiveness of one's own existing QA system. Much of the success of QA measures depends largely on internal reviews, as the organization knows the most pressing issues and concerns—something that is not available in standardized QA systems and indicators. In addition, being transparent and honest about one's own limitations and admitting there are flaws in the delivery system is a facilitating factor for continuous quality improvement. It is the first and most critical step in moving ahead in the QA challenge.

Two issues faced by OUM were discussed in this chapter; that of the supply of learning modules and another related to assignments. The future challenges facing OUM in relation to those issues are as follows:

- There is a pressing need to develop and maintain a robust learner management system to cater to the expanding requirements of the institution. While an increasingly larger number of learners in the system is a plus to the institution in terms of institutional growth and income, it also presents challenges, as every semester the volume of transactions processed (from registration to payment, modules, appointment of tutors/facilitators, examinations questions, and answer scripts) becomes increasingly voluminous. The fact that the learners, tutors/facilitators, and learning centers are widely distributed throughout various geographical locations in the country is a reminder that the University needs to invest heavily in technology to help it stay ahead in the quest to be the leading provider of flexible education.

- Equally important is the need to develop and maintain a robust quality management system, and to ensure that a quality culture is inculcated among staff across all levels of administration. Measures need to be taken to ensure that the University's quality manual, as well as all departments' service operation manuals and standard operating procedures, remain current and are able to bring about desired levels of quality for all learners regardless of geographical locations and academic programs pursued. Further, regular audits must be carried out to determine compliance with established SOPs as well as to identify opportunities for improvement.
- Finally, the greater challenge ahead is to create an environment that encourages innovative and "outside the box" thinking among all staff. This can bring about positive and proactive changes to current and well-established practices and will raise QA standards to greater heights.

Concluding Remarks: Future Policy Directions

Insung Jung

This book has aimed to introduce best quality assurance (QA) practice in distance education (DE), drawing on the experience and knowledge of 16 selected Asian DE institutions. It has discussed the QA policies, frameworks, guidelines, and strategies of the selected DE providers across 12 countries and one territory in Asia. In doing so, the authors have identified the challenges in ensuring and improving quality at an institutional level. The book has also presented the lessons and future directions offered by the Asian DE institutions. It has shown that QA in DE needs to consider a balanced systemic approach, focus on learner support, pedagogical and management aspects, move toward a more performance-based approach, and promote a culture of quality and continuous improvement. In particular, the chapters have revealed that QA will work only if there is commitment from top management, and understanding and acceptance of its strategic directions among all the staff. Quality needs to be fully integrated into the planning and operations of every part of the institution, and into the competencies of all members. Otherwise QA becomes just a "tick-the-box" exercise and slowly fades away. In addition, some of the chapters have indicated that applying QA procedures and reporting systems, establishing QA units, training staff, providing quality learner support, collecting extensive data, and other QA activities bring financial burden to the institutions. This indicates that DE providers should keep exploring more cost-effective ways of conducting QA. At the same time it is important to remember that failure costs resulting from mechanisms, products, or services not conforming to QA requirements or ignoring QA considerations are often higher than the costs of QA itself.

ADOPTING A BALANCED, SYSTEMIC APPROACH

In the book, institutional QA mechanisms have been shown to be strongly affected by the notions of self-improvement and public accountability, which are particularly relevant in Asian DE where hostility to the methods of DE still exist (Daniel 2010). A closer look at the internal QA mechanisms in the DE institutions in Asia has also revealed that tension exists between self-improvement and accountability, often putting strain on the capability of institutions to carry out their internal QA procedures effectively, as argued by Inglis (2005). The authors have described how their institutions manage to meet this challenge. For example, Thailand's Sukhothai Thammathirat Open University has established a QA mechanism that integrates both internal and external QA requirements. As shown in Chapter 2, while both internal and external QA activities overlap, the internal QA unit focuses more on internal quality enhancement and the external QA unit on meeting national QA standards to improve public accountability. In the case of Korea's Hanyang Cyber University (HYCU), all QA standards specified by the government have been deliberately addressed in its internal QA system. In addition, human and physical resources of a well-known conventional university (Hanyang University) have been shared with HYCU in developing the online courses and providing learner services, which has contributed to improving the public image of HYCU (see Chapter 4).

These cases draw some useful lessons for Asian DE providers who wish to develop a balanced QA mechanism addressing both enhancement and accountability.

- It is important for a DE institution to firmly integrate QA standards and criteria set by governments or external assessors into its internal QA processes in such a way as to contextualize national or societal priorities and to reflect unique features of distance teaching and learning.
- It is also important to adopt a comprehensive and systemic approach to QA, and to standardize essential QA measures. QA should begin by identifying or confirming the needs of learners and society by engaging all the stakeholders. Based on the needs, the QA policies, procedures, resources,

implementation strategies, and performance indicators critical to achieving the QA outcomes should be determined.
Then both quantitative and qualitative data should be collected and analyzed for the assessment of the QA outcomes
and further quality enhancement.

- Experienced leadership should be sought to build consensus within the institution and introduce a strong QA
system. Leadership at the top level has to act as a change
agent to develop a quality culture within the institution.
- To carry cut a balanced, integrated, and systemic QA
approach, a centralized QA unit or a QA team that oversees
an institution's overall QA activities and links these activities to the external QA system should be in place.

FOCUSING ON PEDAGOGY, LEARNER SUPPORT AND MANAGEMENT

Quality in pedagogical dimensions such as program and course
development, teaching and learning, evaluation, and learner
support appears to be of great concern to many Asian DE providers and learners, as evidenced in previous studies, including
Baggaley and Belawati (2010), and Jung (2011a).

The best practices shown in this book have emphasized the
importance of detailed QA guidelines for the development and
delivery of programs, courses, and materials. Cases such as the
Indira Gandhi National Open University (see Chapter 11), the
Open University of China (see Chapter 10) and the University of
the Philippines-Open University (see Chapter 12) adopt a
Course Team approach and/or step-by-step course development
processes, and engage external experts for the design and
development of their courses and programs to ensure the quality of
distance teaching and learning materials. Some institutions, such
as Kumamoto's Online Graduate School (see Chapter 9), apply
well-established instructional design models and innovative
teaching and learning paradigms to improve the effectiveness of
its online learning and meet the expectations of its online learners.

In addition to QA measures during the creation and distribution of DE products, QA procedures for the provision of flexible
and satisfactory learner support are well-articulated in our best

practices. As Brindley, Walti, and Zawacki-Richter (2004) show in their edited book on learner support, coordinated actions of faculty, students, administrators, tutors, and learner support practitioners to meet the needs of distance learners, while considering constraints in time and resources, are critical for offering learner-centered yet institutionally bearable learner support services. As presented in Chapter 13, the case of Malaysia's Wawasan Open University challenges the traditional notion of learner support by focusing on the training of human resources. All staff members, including part-time tutors, are key elements of the quality learner support system and are provided with continuous training with well-defined responsibilities, and are rigorously managed in their performance. Further, the Virtual University of Pakistan (VUP) (see Chapter 14), as an e-learning provider to adult learners across over one hundred and eighty campuses in more than 95 cities in Pakistan, provides an example of innovative flexible learner support in its online examination system, which is designed in a secure, convenient, efficient, and reliable manner.

Process-based quality management appears to be another important part of QA measures in DE. In a quality management system, every step of the product development process is monitored, in order to ensure the implementation of QA standards. By adopting ISO's quality management procedures, China's Peking University (see Chapter 6) and Indonesia's Universitas Terbuka (see Chapter 5) seek to maintain and improve the quality of their QA process and products, especially in the areas of material production and delivery, administration and learner services, and organizational change. Similarly, Mongolian e-Knowledge (see Chapter 7) manages the quality of its e-learning development and delivery processes and products by benchmarking a European quality improvement scheme for e-learning. Korea's e-learning company, AutoEverSystems (see Chapter 8), applies its own process management tools at each stage of e-learning development and implementation.

The experiences in these cases provide practical lessons for Asian DE institutions that are in the process of establishing and enhancing their QA system.

- A QA system in a DE institution should safeguard the quality of pedagogical aspects of DE in developing and delivering courses and materials. This can be achieved by adopting a Course Team approach, specifying courses/material development procedures, involving both internal and external experts, considering learners' needs and feedback, and continuously exploring and trialing new learning paradigms and design ideas.

- To ensure and enhance the quality of learner support, it is important to continuously and regularly train faculty, administrators, tutors, and other learner support practitioners, and to relate the quality of their learner support services to performance evaluation.
- A learner support system in any DE institution should focus on needs-based, personalized and flexible learner support, as distance learners have diverse learning needs, and often study under great time and space constraints.
- A DE institution should move from an ad hoc management system to a strong, process-centric QA system in order to achieve the consistency in the quality of its products and services. Already existing quality management systems, such as ISO certification, can be adopted or adapted for this purpose.

MOVING TOWARD A PERFORMANCE- AND OUTCOMES-BASED APPROACH

Our cases show that a DE institution with a strong QA system develops a tight link between its QA efforts and performance measurement. In the case of Open University of Sri Lanka (see Chapter 15), performance indicators specified for each quality standard are used to conduct a self-assessment for continuous quality improvement. These indicators, along with measurable evidence, make it possible for an institution to evaluate the success of its particular QA activity as it determines the role of DE within the larger context of performance improvement, as argued by Rosenberg (2006).

In the cases of Wawasan Open University and the Open University Malaysia (see Chapter 16), we observe a move from a process-based style of QA to a more outcomes-based approach. The performance of various departments and programs is evaluated against the planned outcomes, and the performance of faculty, tutors, administrative staff, and others is assessed annually, based on performance indicators that are created internally or adapted from the Commonwealth of Learning (COL) Performance Indicators for Distance Higher Education institutions.

OECD's *Education at a Glance*[1] proposes to consider both educational and social outcomes in assessing the quality of education. While our cases focus on linking QA efforts and staff performance in the QA system, they do not pay much attention to comparing a wide range of educational outcomes, for example, students' learning performance and social outcomes, such as the impact of education on earnings, employment, and promotion. Also, the cases in our book do not place a high emphasis on whether their students have actually developed the pre-specified competencies, argued as the most important quality indicator in DE by some quality assessors, including the Higher Education Funding Council of England (HEFCE 2005).

As argued in Dabbagh (2007), the quality of DE depends not only on the providers, but also on the capabilities or competencies of the learners and the level and extent of their participation in the learning process. It is thus crucial that DE providers help their learners develop these competencies for the provision of quality DE. Several scholars have identified core competencies of successful distance learners. They include:

- Self-directedness, self-motivation, and self-discipline (Garrison 2003);
- Collaborative, interpersonal, social, reflective, and discursive skills (Anderson 2008);
- Positive attitude toward technology (Schrum and Hong 2002);
- Solid technical skills (Osika and Sharp 2002);
- Time management and scheduling skills (Golladay, Prybutok, and Huff 2000);
- Meta-cognitive competencies (Hong and Jung 2011); and
- Ability to concentrate on study at set times and location (Hong and Jung 2011).

---

[1] http://www.oecd.org/document/52/0,3343,en_2649_39263238_45897844_1_1_1_1,00.html

Jung and Latchem (2011) observe that a performance- or outcomes-based approach to QA is yet to be explored by DE institutions in Asia and other regions. As the governments, employers, the general public, and students demand strong evidences for the quality of DE in Asia, DE institutions need to give more attention to:

- Improving retention and graduation rates;
- Supporting employability and career enhancement;
- Improving satisfaction of learners, employers, and the public with their learning processes and outcomes; and
- Increasing the value and reputation of DE institutions in society.

PROMOTING A CULTURE OF QUALITY AND CONTINUOUS IMPROVEMENT

Harvey and Green (1993) opine that quality is closely connected to an institution's culture. It is important for a DE institution to create and strengthen a quality culture within the institution in order to integrate QA and enhancement activities in everyday practices.

The European University Association (EUA 2006) suggests that the quality culture approach should be development-oriented and value-based, and involve multiple internal and external stakeholders rather than being top-down and controlled. A culture of quality often begins with the leaders and managers. However, QA led or controlled by a small number of top managers or the government without shared beliefs, values, attitudes, and behavior patterns within the organization results in only conformity to external QA standards and seldom contributes to continuous quality improvement. So the foremost task in developing the internal quality culture is to develop common understanding and shared sense of quality and QA, as there are different understandings of quality and QA by different stakeholders, who hold a variety of values and ideals toward DE. The next step would be, based on shared understanding of quality and QA, to make the quality culture more formal and explicit.

Our cases provide important benchmarking points for Asian DE institutions in developing an internal quality culture for continuous quality improvement.

- The vision, purpose, and mission of the institution have to be shared and clear, both within the institution and to external actors, and reflect institutional values for improvement of DE products and services.
- Leadership and top management should support the quality culture for change and transformation, as evidenced in several cases in our book, including the cases of Singapore's UniSIM (see Chapter 1) and the Virtual University of Pakistan.
- The institution should not focus on inspection or assessment but build quality into its products and services from the start.
- The institution should engage every member in QA activities and in accomplishing the transformation, as shown in the cases of Open University of Hong Kong (see Chapter 3) and Indonesia's Universitas Terbuka. Introducing a special QA unit does not guarantee the actual quality improvement unless quality product and process ownership is resident at all levels and units of the institution.
- The institution should build a strong and continuous self-improvement system for all the members.
- The institution should encourage collaborative teamwork, benchmarking innovative practices, exploring new ideas as well as quality performance. Its QA system should not be operated too rigidly or managerially.
- All stakeholder evaluation should be fed into the continuous quality improvement cycle.

CONCLUSION

Not all governments have developed a national policy framework for QA in DE, and not all societies have shown favorable attitudes toward the quality of DE. In spite of such often unfavorable conditions, several Asian institutions have developed policies and strategies for improving the quality of their teaching and learning and achieved the public trust and satisfaction of their stakeholders. Nevertheless, there are still many DE institutions in Asia that do not pay enough attention to the quality issues, or apply piecemeal approaches to address problems related to the quality in DE. In building and improving the QA and enhancement systems and policies, these institutions will benefit from benchmarking the best practices offered in this book and from considering two issues discussed below.

Asian Learners' Perception of Quality in Distance Education

While inputs from the providers, assessors, and governments are valuable in examining and promoting quality in DE (Frydenberg 2002), it is also critical to understand learners' views. The success of DE does not simply derive from the products and services delivered to the learner, but from the knowledge, understanding, and relationships that are co-developed by both learners and teachers during the teaching and learning processes, and typically relies to a great extent on learners' motivation and engagement (Jung 2011a).

A study conducted with one thousand, six hundred and sixty five Asian DE learners who were enrolled in 24 DE institutions or programs across 11 Asian countries and one territory reveals some unique features of Asian distance learners' perception of quality in DE (Jung 2011b). All ten quality dimensions—Faculty Support, Student Support, Information and Publicity, Course Development, Interactive Tasks, Teaching and Learning, Evaluation and Assessment, Infrastructure, Internal QA Mechanism, and Institutional Credibility—proposed in the study appear to be important in explaining the quality of DE from the Asian learners' perspective. Student Support and Information and Publicity dimensions were quite powerful in measuring the quality of the support aspect, while Faculty Support was less important. The Asian distance learners perceived a DE institution or program that provides social and psychological support and clear course information to be of high-quality.

With respect to pedagogy, Evaluation and Assessment was most powerful in explaining the quality of DE. The learners saw fair and clear learning assessment guidelines and periodic students' evaluation of teaching and learning to be particularly important. Course Development was also influential in evaluating the quality of the academic aspect of DE. Asian learners perceive a DE program that provides well-structured course materials that follow clear development procedures and are considerate of learners' needs to be of high-quality. Other pedagogical aspects such as Teaching and Learning and Interactive Tasks were slightly less important compared with Course Development and Evaluation and Assessment, but still influential in assessing the pedagogical quality of DE.

Infrastructure appeared to be most important in assessing the quality of the environmental aspect of DE. This finding highlights the importance for DE providers to ensure that dependable technology infrastructure is in place. At the same time, it is important to remember that much of Asia is still constrained with limited infrastructure and skills, high costs, and slow internet speeds, and thus the combined use of analogue and digital technologies is highly recommended for Asian DE (Baggaley and Belawati 2010). Internal QA Mechanism and Institutional Credibility also appeared to be influential environmental aspects in assessing the quality of DE. It was also found that, in the eyes of the Asian learners, both institutionalization of an internal QA system under strong leadership and recognition by external accreditation agencies are important for gaining public trust in DE.

These results suggest that Asian DE providers should consider their learners' perspective in developing and improving the QA system, since their views reveal important quality areas that are not always reflected in the assessors' QA guidelines.

Gender Issues

In Asia, gender disparity is a serious issue in education. UNICEF (2009) reports that although steady progress has been made in achieving some gender parity regarding secondary enrolment ratios, the situation is still far from satisfactory, and that, while female enrolment in higher education has increased globally, it is not the case in most parts of Asia. As Baden and Green (1994) argue, the gender disparity in education in Asia can be explained by several constraints resulting from historical, economic, socio-cultural, and school-related factors. Historical factors are related to national policies, which place emphasis on a campus-based education and an elite education Economic factors refer to society's low employment rate, wage discrimination against women, and the tendency to allocate household resources to men first. Socio-cultural factors that affect gender disparity in education include religion, age at marriage, marital system, and ethnicity. School-related factors involve the school types, locations, curriculum bias, and less representation of female teachers in schools.

DE has emerged as a valuable means to overcome some of the constraints caused by these factors and to contribute to widening access to education and reducing the gender disparity in education. Over half of the learners enrolled in the Asian DE institutions are female. In this regard, we can say that DE has expanded the opportunity for women and girls to access higher education. However, there has been a growing public concern over the quality of DE delivered and high dropout levels, especially among females.

As discussed above, quality is a relative, value-laden concept and there may be differences in the views of quality between male and females. Green and Trevor-Deutsch (2002) observe that female students in Asian DE face barriers when the course content is not directly relevant to their livelihood; when it does not value their knowledge, wisdom, and experience; when access to the content is too costly; and when they do not feel able to use the technology competently or confidently. This is supported by case studies that detailed how Asian female distance learners had overcome frustrations and succeeded in their learning (Kanwar and Taplin 2001). As Taplin and Jegede (2001) and Von Prümmer (2000) argue, Asian female learners ask for support that assists them to overcome personal and social barriers and achieve high performance. In the study mentioned earlier, Jung and Fukuda (2011) report gender differences in the perceived problems with, and concerns over, DE. "Conflict with family obligation" was the most serious barrier for the female learners, whereas "financial difficulties" was prominent for their male counterparts.

These studies suggest that Asian DE providers should understand that, for female learners, quality DE means a system that removes these barriers, that maximizes opportunity, that provides needs-based learner supports, and that is based on the understanding of their perceptions, concerns, and experiences. The providers should thus consider these gender differences when designing a QA system.

REFERENCES

Anderson, T. 2008. "Teaching in an Online Learning Context," in T. Anderson (ed.), *Theory and Practice of Online Learning* (2nd Ed.), pp 343–366. Canada: Athabasca University.

Baden, S. and Green, C. 1994.*Gender and Education in Asia and the Pacific*. Report submitted to the Australian International Development Assistance Bureau. BRIDGE (Development–Gender) Institute of Development Studies University of Sussex.

Baggaley, J. and Belawati, T. 2010. *Distance Education Technologies in Asia*. India: New Delhi: SAGE Publications India Pvt. Ltd.

Brindley, J.E., Walti, C., and Zawacki-Richter, O. (eds). 2004. *Learner Support in Open, Distance and Online Learning Environments*. Oldenburg: Bibliotheks- und Informationssystem der Universität Oldenburg.

Daniel, J. 2010, October. "Distance Education under Threat: an Opportunity?" Keynote address at the IDOL & ICEM 2010 Joint Conference and Media Days in Eskisehir, Turkey. Available online at http://www.col.org/resources/speeches/2010presentation/Pages/2010-10-06.aspx and accessed on September 4, 2012.

Dabbagh, N. 2007. "The Online Learner: Characteristics and Pedagogical Implications,"*Contemporary Issues in Technology and Teacher Education*, 7(3), 217–226.

European University Association (EUA) 2006. *Quality Culture in European Universities: A Bottom-up Approach*. Brussels: EUA.

Frydenberg, J. 2002. "Quality Standards in e-Learning: A Matrix of Analysis," *International Review of Research in Open and Distance Learning*, 3(2). Available online at http://www.irrDE.org/index.php/irrDE/article/viewArticle/109/189 and accessed on September 4, 2012.

Garrison, D. R. 2003. "Self-Directed Learning and Distance Education," in M. G. Moore and W. Anderson (eds), *Handbook of Distance Education*, pp. 161–168. Mahwah, NJ: Lawrence Erlbaum.

Golladay, R., Prybutok, V., and Huff, R. 2000. "Critical Success Factors for the Online Learner,"*Journal of Computer Information Systems*, 40(4), 69–71.

Green, L. and Trevor-Deutsch, L. 2002. *Women and ICTs for Open and Distance Learning: Some Experiences and Strategies*. Vancouver: Commonwealth of Learning. Available online at http://www.col.

org/SiteCollectionDocuments/women%20and%20ICTs.pdf and accessed on September 4, 2012.

Harvey, L. and Green, D. 1993. "Defining Quality," *Assessment and Evaluation in Higher Education*, 18(1), 9–34.

HEFCE (Higher Education Funding Council of England). 2005. *HEFCE Strategy for e-Learning*. Available online at http://www.hefce.ac.uk/pubs/hefce/2005/05_12/ and accessed on September 4, 2012.

Hong, S. and Jung, I.S. 2011. "The Distance Learner Competencies: A three-Phased Empirical Approach," *Educational Technology Research and Development*, 59(1), 21–42.

Inglis, A. 2005. "Quality Improvement, Quality Assurance, and Benchmarking: Comparing Two Frameworks for Managing Quality Processes in Open and Distance Learning," *The International Review of Research in Open and Distance Learning*, 6(1). Available online at http://www.irrDE.org/index.php/irrDE/article/view/221/304 and accessed on September 4, 2012.

Jung, I.S. 2011a. "The Dimensions of e-Learning Quality: From the Learner's Perspective," *Educational Technology Research and Development*, 59(4), 445–464.

Jung, I.S. 2011b. *Asian Learners' Perception of Quality in Distance Education*. Report submitted to the Virtual University of Pakistan as a technical report for the Openness and Quality in Asian Distance Education Project funded by the International Development Research Centre (IDRC) of Canada.

Jung, I.S. and Fukuda, A. 2011. "Gender Differences in Asian Learners' Perception of the Quality in Distance Education and e-Learning: Implications for a Gender-Considerate Support System," Paper presented at the 25th AAOU conference, Penang, Malaysia, September 28–October 1.

Jung, I.S. and Latchem, C. 2011. "Concluding Remarks: Quality matters," in I.S. Jung and C. Latchem (eds), *Quality Assurance and Accreditation in Distance Education and e-Learning: Models, Policies and Research*, pp. XX–XX. New York and London: Routledge.

Kanwar, A.S. andTaplin, M. 2001. (eds). *Brave New Women of Asia: How Distance Education Changed their Lives*. Vancouver: The Commonwealth of Learning.

Osika, E. R. and Sharp, D. P. 2002. "Minimum Technical Competencies for Distance Learning Students," *Journal of Research on Technology in Education*, 34(3), 318–325.

Rosenberg, M.J. 2006. *Beyond e-Learning: Approaches and Technologies to Enhance Organizational Knowledge, Learning, and Performance*. New York: Pfeiffer.

Schrum, L. and Hong, S. 2002. "Dimension and Strategies for Online Success: Voices from Experienced Education," *Journal for Asynchronous Learning Networks*, 6(1), 57–67.

Taplin, M. and Jegede, O. 2001. "Gender Differences in Factors in Influencing Achievement of Distance Education Students," *Journal of Open, Distance and e-Learning*, 16(2), 133–154.

UNICEF. 2009. *Towards Gender Equality in Education: Progress and Challenges in Asia-Pacific Region*. Retrieved from http://www.ungei.org/resources/files/Towards_Gender_Equality_in_Education_051809.pdf and accessed on September 4, 2012.

Von Prümmer, C. 2000. *Women and Distance Education: Challenges and Opportunities*. New York: Taylor and Francis Group.

About the Editors and Contributors

EDITORS

Insung Jung is currently Professor of Education at the International Christian University (ICU) in Tokyo, Japan. Before joining ICU in 2003, she served as the Director of the Multimedia Education Center at the Ewha Women's University, Seoul and was on the faculty of the Korea National Open University, Seoul. She has also served as a consultant and technical advisor in distance learning to numerous national and international institutions, including UNESCO's Open and Distance Learning Initiative for Higher Education Knowledge Base, the APEC ICT Human Capacity-Building and Facilitation of Human Resources Exchange, and the Advisory Committee of World Bank's GDLN project in Korea. Her recent publications include *Distance and Blended Learning in Asia* and *Quality Assurance and Accreditation in Distance Education and e-Learning*.

Tat Meng Wong was formerly a Professor of Wawasan Open University (WOU), and also the former Vice Chancellor and CEO of WOU Penang. He also served as the President of Asian Association of Open Universities (AAOU) from January 2011 to September 2012. He graduated with a PhD in Zoology from Otago University in New Zealand in 1972 and has more than 40 years experience in conventional and ODL institutions, having held senior positions in Universiti Sains Malaysia (1972–1993), Open University of Hong Kong (1993–2006), and WOU (from 2006–September 2012). His current interests include QA in ODL systems and change management.

Tian Belawati is Professor and currently Rector of Universitas Terbuka (Indonesia Open University). She has been working in the field of distance education since 1985 and has presented many papers at national and international conferences, and has published numerous books and academic journals. She has served as Secretary General (2008–2009) and President (2009–2010) of the Asian Association of Open Universities (AAOU) and is President of the International Council for Open and Distance Education (ICDE). She has also been actively involved in research on the use of ICT for education in Asia with the PANdora Network, on which she now serves as a Board Member.

CONTRIBUTORS

Anuwar Ali is President/Vice Chancellor and Professor Emeritus of the Open University Malaysia (OUM). Prior to the current appointment, he was the Vice Chancellor of Universiti Kebangsaan Malaysia (UKM) for a period of five years (1998–2003). Professor Emeritus Anuwar Ali obtained his PhD from the University of Kent, Canterbury, in 1982. During his academic career, he has held various posts including the Director of Higher Education, Ministry of Higher Education (1996–1998). He also served as chairman as well as board member of a number of institutions including the Securities Commission Malaysia, the Malaysian Examinations Council, and the International Medical University. He has also lent his expertise to the Malaysian government and international bodies such as UNIDO and UNDP over the years.

Patricia B. Arinto is Dean of the Faculty of Education at the University of the Philippines Open University (UPOU), where she also teaches courses on foundations of distance education, learning theories and instructional design, and online teaching and learning. She has extensive experience in distance education course and program development, and the design of teacher professional development programs in technology integration in the curriculum in the Philippines and other parts of Asia.

Sanjaa Baigaltugs is Professor in the Computer Engineering Department of the School of Computer Science and Management at the Mongolian University of Science and Technology. His current research focuses on the distribution and effective use of e-learning platforms in Mongolian universities. Professor Baigaltugs has published over 30 professional papers and research articles on topics including quality assurance mechanisms in higher education and the implementation of credit systems to the higher education sector in Mongolia.

Niu Ben is a staff member at the International Cooperation and Exchange Division of the Open University of China. He participated in the project "Research on the Construction of Open Universities in China," funded by the Ministry of Education in 2010. His academic interest focuses on management of open and distance education.

Theppasak Boonyarataphan is an Associate Professor in the School of Management Science at STOU, and has served as Vice President for Planning and System Development. He earned his PhD from the National Institute of Development Administration. At STOU, He has developed and administered policies on quality assurance (QA). He has also served as a QA assessor of higher education institutions throughout Thailand. As an expert in public management science, Dr Theppasak has been invited by public and private organizations to give advice on matters of modern administrative planning and developing information systems to support organizational administration both in Thailand and overseas.

Robert Edward Butcher is currently the Executive Coordinator and Secretary to Council of the Open University of Hong Kong. He was educated in the UK, obtaining his Bachelor's degree from University College London and his PhD from the University of East Anglia. Bob has over 25 years experience working in the tertiary education sector, mostly in universities dedicated to adult, distance, and open learning. Before relocating to Hong Kong in 1993, he spent five years working with the Open University in the UK where he was responsible for the policies and procedures related to tuition and counseling. His areas of expertise are in teaching and learning strategies and quality assurance; he has undertaken several international consultancies in places such as South Africa, Malaysia, and Lesotho.

Uma Coomaraswamy is Emeritus Professor of Botany at The Open University of Sri Lanka. She is an Honorary Fellow of the Commonwealth of Learning and functions as a Consultant to Lifelong Learning for Farmers (L3F). In this role, she has implemented L3F Projects using Technology Mediated Open and Distance Education at three national universities in Sri Lanka. Her university career spans over 40 years at the University of Colombo, Eastern University, and the Open University. She was the Vice-Chancellor of the Open University of Sri Lanka and, after retirement, consultant on Quality Assurance in the ADB/ Distance Education Modernization Project for three years, during which period she developed QA tools, systems, and processes for distance higher education at a national level. She also spearheaded the preparation of the "Quality Assurance Toolkit for Distance Higher Education Institutions and Programs" published by COL in 2009.

Mansor Bin Fadzil is Senior Vice President and Professor at the Open University Malaysia. He obtained a PhD in Control System Engineering from the University of Sheffield in 1985, after which he served at the University of Malaya (UM) until 2001. During his tenure at UM, Professor Mansor held various posts, including the Director of the Distance Learning Center, Special Assistant to the Vice-Chancellor, and Director of Multimedia Development Center.

Sri Y.P.K. Hardini has been an academic staff in the Faculty of Mathematics and Natural Sciences, Universitas Terbuka (UT) in Jakarta, Indonesia, since 1989. She is responsible for course material and item test development, especially for poultry management and animal husbandry in distance learning. She has many years of experience in distance education in Indonesia, and she was Archieve Module Coordinator in the Multi Media Production Center, UT from 2004–2008. Since 2008 she has been Head of the Quality Assurance Center, after being a Coordinator of Quality Control in the Center from 2005–2008.

Yeonwook Im is an Associate Professor in the Department of Educational Technology and Dean of the Office of International Affairs at Hanyang Cyber University in Seoul, Korea. She received her Bachelor's degree at Seoul National University, her Master's degree at Harvard University, and a doctoral degree at the University of Pittsburgh. She is currently a member of the policy counseling committee for the Ministry of Education, Technology, and Science in Korea. She has worked for the establishment and development of cyber universities for 10 years. Her research interests include online learning, distance education, instructional design, blended learning, new media in education, and e-learning quality assurance.

Cheong Hee Kiat is the founding President and Professor of SIM University (UniSIM). He was Deputy President and Dean of Civil and Environmental Engineering at Nanyang Technological University before joining UniSIM in 2005. In academia since 1986, his experience covers teaching, research, international relations, leadership, and administration. He has served on various education and public committees and boards, and has been involved in university academic audits in Singapore, Australia, Hong Kong, Korea, and Ireland. He chairs the External Audit Panel of the Polytechnic Quality Assurance Framework and is a member of the Engineering Accreditation Board in Singapore. Professor Cheong studied Civil Engineering at the University of Adelaide and obtained his MSc and PhD from Imperial College, London.

Chen Li is Professor of Educational Technology at Beijing Normal University, Director of the Research Center of Distance Education, and Executive Dean at the Beijing Institution for Lifelong Society. She also manages the Master's degree program in distance education at Beijing Normal University. Chen Li has been researching interaction theory and interactive media, and has published more than 10 papers on these topics since 2004. Chen Li has been involved in many international collaborative projects.

Teik Kooi Liew is currently the Quality Assurance Manager at Wawasan Open University (WOU). He joined WOU as Assistant Registrar in 2005, prior to its formal establishment in August 2006, and was promoted to Senior Assistant Registrar in 2008. He graduated with a PhD in Biotechnology (Plant Cloning) from Universiti Sains Malaysia in 2000. Prior to joining WOU,

Dr Liew had held various academic and administrative positions, including Special Functionary to the Principal and Head of the Pre-University Studies Department in Disted Stamford College (2000–2003), Head of the Center of Pre-University Studies, KDU College Penang (2004–2005), and Administration Manager, Allianze College of Medical Sciences (currently Allianze University College of Medical Sciences) (2005–2006).

Cheolil Lim is Professor in the Educational Technology Program at Seoul National University. He received his PhD from Indiana University in 1994. His research is in the areas of instructional systems design, learning environment design, distance education, corporate education, and e-learning. Dr Lim is former Vice President of the Korean Society for Educational Technology and former President of the Korean Society for Learning and Performance. He has served on the editorial boards of many key Korean journals. His current projects focus on designing support systems for creativity, self-regulated learning, and learning design.

Naveed Akhtar Malik is Rector and Professor at the Virtual University of Pakistan. He obtained his Master's degree in Physics from the University of the Punjab, Pakistan, and a Doctor of Science degree in Electrical Engineering from M.I.T., USA. He is the founder of the Virtual University of Pakistan, and is the principal architect of its highly innovative and acclaimed ICT-based e-assessment system. Dr Malik is currently coordinating an IDRC (Canada) funded Asia-wide research effort focused on openness and quality in Asian distance education. He is a life member of the Pakistan Institute of Physics, and was awarded the national award of Sitara-e-Imtiaz in 2008 for his services to education.

Any Meilani has been an academic in the Management Study Program, Faculty of Economics, Universitas Terbuka (UT) in Jakarta, Indonesia, since 1989. She is responsible for course material and item test development, especially for Management in distance learning. She has many years of experience in distance education in Indonesia, and she was head of the Management Study Program during 2004–2006. She was also responsible for the Item Bank in the Faculty of Economics during 2001–2004.

Since 2009 she has been a Coordinator of Performance Appraisal in the Quality Assurance Center. Her research interests include quality assurance implementation, especially online tutorial in distance education, and Syariah Economics and Finance.

Rose Nembiakkim is currently an Associate Professor in the School of Social Work at the Indira Gandhi National Open University (IGNOU), New Delhi. She has also worked in IGNOU's Staff Training and Research Institute of Distance Education, and has published several papers in ODL.

Pema Eden Samdup is currently an Assistant Professor in the School of Humanities at the Indira Gandhi National Open University (IGNOU), New Delhi. Formerly, she has worked as Program Officer (Open Schooling and Higher Education) with the Commonwealth Educational Media Center for Asia, New Delhi. She is deeply interested in using ICT tools for the humanities, and has ODL experience of nearly 10 years. She has several published papers in ICT and ODL, as well as in her discipline.

Gao Shuping is Executive Dean at Peking University's School of Distance Learning for Medical Education.

Hae-Deok Song is Associate Professor of Education and Director of the Center for Teaching and Learning at Chung-Ang University, South Korea. Prior to that, he worked as an assistant professor of education at State University of New York, Albany, USA. He received his PhD from Penn State University in 2004. His research is in the areas of problem-based learning, the adoption of innovative technologies in higher education settings, and pedagogical user interface design principles.

Deetje Sunarsih has been an academic staff member in the Chemistry Education Study Program, Faculty of Teacher Training and Educational Sciences, Universitas Terbuka (UT) in Jakarta, Indonesia, since 1935. She is responsible for course material and item test development, especially for Chemistry Education in distance learning. She has many years of experience in distance education in Indonesia, and was head of the Chemistry Education Study Program (1995–1997), and head of UT Bogor Regional

Office during 2004–2008. Since 2008 she has been a coordinator of Quality Control in the Quality Assurance Center. Her research interests include quality assurance implementation, especially in distance education and environmental education.

Pranee Sungkatavat is President of Thailand's Sukhothai Thammathirat Open University (STOU). She held several senior administrative positions including Assistant to the President, Vice President for Educational Services, and Vice President for Academic Affairs at STOU before becoming President in 2008. After obtaining two Master's degrees in the US, she received a PhD in Curriculum and Instruction from Kansas State University. Dr Pranee has also served as an assessor and evaluator of the quality of education curricula in Thai higher education institutions. Her publications include: "The Development of Distance Education Curricula", "The Development of Non-Formal Education Curricula, and Curriculum Design". In 2008, Dr Pranee was awarded a distinguished alumnus of the Faculty of Education, Chulalongkorn University, and designated as a representative of Thailand to the SEAMEO Regional Open Learning Center Governing Board.

Katsuaki Suzuki is Professor and Chair of the Graduate School of Instructional Systems at Kumamoto University. He earned his PhD, in Instructional Systems from Florida State University, USA. Currently, he is a director of the International Board for Standards of Training, Performance and Instruction (ibstpi®), a contributing editor for Educational Technology Research and Development, and Coeditor-in-Chief of the International Journal of Media in Education, jointly published by the Japan Association for Educational Media Study and the Korean Association for Educational Information and Media. He also serves as a board member of the Japan Society for Educational Technology, Japanese Society for Information and Systems in Education, and Japan Society of Instructional Systems in Healthcare, as well as an honorary member of e-Learning Consortium Japan.

Yang Tingting is Senior Researcher in the Research Institute of Open and Distance Education at the Open University of China, and has conducted research on open and distance education for over 30 years. She has obtained many awards at the national and provincial levels for her research projects, mainly in educational technology and quality management of open and distance education. In 2008, she published a monograph on quality assurance in open and distance education with respect to OUC's academic quality assurance system.

Shen Xinyi is a Master's student in the Research Center of Distance Education at Beijing Normal University.

Li Yawan is Professor in the International Cooperation and Exchange Division of the Open University of China. In 1997, as a visiting scholar, she undertook research on open learning systems in the UK, supported by the Ministry of Education. Between 2005 and 2007, she was Secretary General of AAOU. Her research areas focus on comparative studies of student support, quality assurance, and other topical issues in the field of open and distance education. She has published research articles in both Chinese distance education periodicals and international distance education journals. She promotes collaboration between the Open University of China and other open and distance education institutions, in order to enhance mutual understanding in a cross-culture context.

Liu Yiguang is President, Mycourse Online Education Technology Co. Ltd., and Pre-assistant Dean, Peking University's School of Distance Learning for Medical Education.

Index

academic bodies overlooking QA, at
IGNOU, 178
academic performance audits (APA) of
HEIs, by MQA, 200
academic QA framework, of UNiSIM,
9–11
accreditation agencies/councils,
establishment of, xxii
Accrediting Association of Chartered
Colleges and Universities of the
Philippines (AACCUP), 183
Advisory Peer Group (APG), 46
agency, results of QAAS by, 141
All India Council of Technical
Education (AICTE), 170
annual enrollment, 60
ASEAN Quality Assurance Network,
xxii
Asia e University, Malaysia, 200
Asian Development Bank (ADB),
105–6, 243–44
Asian Management Award, to STOU
in 1996, 30
Asian Open Universities (AAOU), 84
Asia Pacific Quality Network (APQN),
xxii
assignment management, quality
assurance in, 213
Associate Faculty (AF), 13, 15
AutoEver (AE), Hyundai
E-learning unit at, 119
individualized learning process
management, 123
quality assurance system and tools
challenges and solutions, 129–32
e-Learning programs and QA

course evaluation systems,
120–23
future development, 133–34
lessons learned, 132–33
organizational structure, 120
during program development
stage, 123–26
during program implementation
stage, 126–28
recognition by Ministry of Labor,
119
automobile-specific programs, of AE
e-learning program, 122
Autonomous Universities (AUs), 3–4

Badan Akreditasi Nasional Perguruan
Tinggi (BAN-PT). *See* National
Accreditation Board for Higher
Education, Indonesia
Badan Standar Nasional Pendidikan
(BSNP). *See* National Board
of Education Standardization,
Indonesia
balanced scorecard (BSC) system, at
OUM, 264
benchmarking system, UNiSIM with
other universities, 17
Business Breakthrough University,
Japan, 142

China Central Radio, 155. *See also*,
Open University of China
(OUC)
class size, 52, 73, 75
Code of Practice for Institutional Audit
(COPIA), Malaysia, 259

Code of Practice for Program
  Accreditation (COPPA),
  Malaysia, 259
collaborative programs, at OUHK, 54
Commission on Higher Education
  (CHED), Philippines, 183–85
Committee of Vice-Chancellors and
  Directors (CVCD), Sri Lanka,
  241
Commonwealth of Learning Review
  and Implementation Model
  (COL-RIM), 245–46
competency-based approach, at GSIS
  for promotion of outcome-based
  QA, 148–49
content development, at OUHK,
  69–71
continuous assessment (CA), 14–15
convoy system, Japan, 140
corporate e-learning, in Korea, 118
correspondence education, xxi
Council for Private Education (CPE),
  4–5, 22
course
  design and implementation policies
    of GSIS, 147–48, 151
  registration procedure at HYCU, 67
course delivery model, of WOU, 204–5
course development, at IGNOU
  quality assurance in, 172–74
Course Production and
  Administration Team (CPAT),
  at STOU, 31
course team meetings, building at
  GSIS, 145–47
credit units (CU), in UNiSIM bachelor
  degree programme, 9
curriculum design, at IGNOU
  quality assurance in, 172–74
customer relationship management
  (CRM) software, at UNiSIM, 16
cyber universities, Korea, 57. *See also*,
  Hanyang Cyber University
  (HYCU), Korea
  profile of, 58–60
  quality assurance in, 60–65
Cyber University Establishment
  Committee, Korea, 61

detailed program proposal (DPP), at
  OUHK, 46
digital library, 62, 204, 226, 244, 262–63
Distance Education Council (DEC),
  India, 178
  accreditation to SOUs, 169–70
  Asian learners' perception of quality
    in, 283–84
  challenges faced by, 169
  development of guidelines for ODL
    programs, 170
  empowerment to IGNOU, 169
  prerequisites for quality in online
    education, 170
distance education (DE), 115–17, 275,
  278–79, 281
  in China, xvii, xxii, 94–95
  development in Mongolia, 106–9
  in Hong Kong, 44
  in Japan, 141
  in Philippines, 182
  Sukhothai Thammathirat Open
    University (STOU), Thailand,
    29, 36
Distance Education Modernization
  Project (DEMP), OUSL, 243–44

educational system model of cyber
  universities, 63
Education at a Glance (OECD), 280
E-Education Project, Mongolia, 108–9
e-learning, xvii, 115–17
  at HYCU, competency of evaluator,
    75–76
  industry in Korea, 118
  in Mongolia, 105
  at OUHK, 53–54
  at UNiSIM, 13–14
E-learning Center, in Mongolia, 107
e-Learning Consortium (eLC), Japan,
  148
e-Learning Industry Development
  Act in 2004, Korea, 118
E-Mongolia National Program, 108.
  See also, ICT Vision 2010
Employment Insurance
  Reimbursement Policy, Korea,
  118

end-of-course assignment (ECA), 14–15
enrollment, annual
  cyber universities of Korea, 60
establishment-approval system (EAS),
  Japan, 139–40, 143
European Foundation for Quality
  Management (EFQM), 110
European Union, 105
European University Association
  (EUA), 281
evaluation criteria for qualitative
  analysis, at Korean cyber
  universities, 62
External Course Assessor (ECA),
  appointment at OUHK, 48
external QA systems for HEIs,
  Thailand, 25, 33–35. *See also,*
  Office for National Education
  Standards and Quality
  Assessment (ONESQA),
  Thailand

face-to-face provision, at OUHK, 51–53
faculty support, QA at OUHK, 73
feedback by students, on UNiSIM
  education system, 16–17
flexible learning, 19–20

gender disparity, in education in Asia,
  284–86
general programs, of AE e-learning
  program, 120–22
grade point average (GPA) system, in
  UNiSIM, 15
graduate programs, at HYCU's, 66
Graduate School of Instructional
  Systems (GSIS) Japan,
  Kumamoto University's
  areas of study, 142
  establishment of, 142
  issues faced and exploration of
    solution, 151–52
  master's program of, 143
  quality assurance measures at
    adoption of competency-based
      approach, 148–49
    built-in course team meetings as
      faculty, 145–47

conducting self- and external
  evaluation, 143–45
establishment approval system
  (EAS), 143
implementation of course design
  and policies, 147–48
introduction of story-centred
  curriculum, 149–50
offering of initial student
  orientation, 150–51

Hanyang Cyber University (HYCU),
  Korea, 276. *See also,* Cyber
  universities, Korea
establishment of, 58
quality assurance at
  issues and challenges, 74–76
  mechanisms, 69–74
  overview of, 65–69
Higher Education Commission (HEC),
  Pakistan, 220
  Accreditation Councils
    establishment, 222
  quality assurance framework
    development, 221
Higher Education Institutions (HEIs),
  25, 58, 82, 84, 105, 258
  in Malaysia, 199
  Malaysian Qualifications Agency
    (MQA), 199, 258–59
  in Thailand, 25
Higher Education Modernization Act
  (Republic Act 8292), 183
Hong Kong Council for the
  Accreditation of Academic
  and Vocational Qualifications
  (HKCAAVQ), 42, 45, 47

i-Campus student management
  system, 267–68
ICT Vision 2010, 107–8. *See also,*
  E-Mongolia National Program
Indira Gandhi National Open
  University (IGNOU), 169, 277
  academic review by DEC in 2007, 170
  establishment of, 171
  national and international operation,
    171

quality assurance a:
  in course development, 175–77
  in curriculum design, 172–74
  guidelines for exporting
    programs, 172
  issues and problems, 177–79
Indonesian Open University. *See*
  Universitas Terbuka (UT),
  Indonesia
industrial production line, xvii
information and communication
  technologies (ICTs), xix, xvii
information technology (IT), 142
Institute of Quality, Research, and
  Innovation (IQRI), Malaysia,
  262–63
institutional accreditation procedure,
  initial at Korean cyber
  universities, 61
instructional design (ID), 142
instructional management (IM), 142
intellectual property (IP), 142
internal QA systems for HEIs,
  Thailand, 25
  components and indicators for, 26
  at STOU, 31–33
Internal Validation Committee (IVC),
  46
International Academic Advisory
  Panel (IAAP), 4
International Development Research
  Centre (IDRC), 106
internet, xvii, xxi
Internet-based Distance Education
  Project, in Mongolia, 106–7
ISO 9000, case study of SDLME
  as a tool to produce organizational
    changes, 101–2
  as a tool to promote process
    analysis, 102–3
ISO 9001 for Quality Management
  System, 87–88

Japan
  distance education existence in,
    141
  quality assurance (QA) framework
    in higher education, 139

Japan Institution for Higher Education
  Evaluation (JIHEE), 140
Japan University Accreditation
  Association (JUAA), 140

Kenichi Ohmae Graduate School of
  Business, Japan, 142
Knowledge Network, 106
Korea Research Institute for
  Vocational Education &
  Training (KRIVET), 118–19

leadership succession, 21–22
learning management system (LMS)
  at HYCU, 65, 67–68
  learner support tools in, 131
learning support services, at OUC,
  160–61
Lembaga Akreditasi Negara (LAN).
  See Malaysian Qualifications
  Agency (MQA)
library, 16, 69, 108, 184, 210, 212

Mainland China, OUHK programs
  in, 53
Malaysian Qualifications Agency
  (MQA), 258
  academic performance audits of
    HEIs, 200
  responsibility for ensuring quality
    of public and private HEIs,
    199–200
Malaysian Qualifications Framework
  (MQF), 199, 259
Malcolm Baldrige National Quality
  Award model, 38–40. *See also,*
  Sukhothai Thammathirat Open
  University (STOU), Thailand
management of examinations, quality
  assurance in, 215
mentoring program, for QA at OUHK,
  74
Ministry of Education, Culture and
  Science, Mongolia (MOECS),
  106–7
Ministry of Education, Culture,
  Sports, Science and Technology
  (MEXT), Japan, 139–40

Ministry of Education (MOE)
China, 94, 155
Singapore, 3–4
Ministry of Education, Science and
Technology (MEST), Korea, 57,
60, 76
mobile learning, 74–75
Mongol Education, 107
Mongolia
distance education development
and NGO involvement, 106–9
e-learning programs in, 105
Mongolian e-Knowledge (MeK) and
quality assurance
issues and challenges, 112–13
mechanisms of, 110–12
overview of, 109–10
as process, 112
Mongolian Foundation for Open
Society (MFOS), 106
Mongolian National Council for
Education Accreditation
(MNCEA), 105
Mongolian University of Science and
Technology (MUST), 107
monitoring criteria and scoring
method, of cyber universities,
64
MyUniSIM learning management
system, at UNiSIM, 13, 16

Nanyang Technological University,
Singapore, 3
National Accreditation Board Act
1996, 258
National Accreditation Board for
Higher Education, Indonesia,
81–82
National Assessment and
Accreditation Council (NAAC),
India, 169–70
National Board of Education
Standardization, Indonesia,
81–82
National Council for Teacher
Education (NCTE), 170
National Education Act, B.E.2542
(1999), 27

National Institution for Academic
Degrees and University
Evaluation (NIAD-UE), Japan,
140–41, 145
National Program on Distance
Education, Mongolia, 108
National University of Singapore, 3

Office for National Education
Standards and Quality
Assessment (ONESQA),
Thailand, 36–37
evaluation framework, standards for
academic services to society,
27–28
institutional and human resources
development, 28
internal QA system, 28–29
preservation of art and culture, 28
quality standard of graduates, 27
research and creative work, 27
second standard, 27
third standard, 27–28
external QA for HEIs, 25
Office of Academic Support and
Instructional Services (OASIS),
Philippines, 187–88
Office of the Higher Education
Commission (OHEC), Thailand,
25
Office of the Vice-Chancellor for
Academic Affairs (OVCAA),
Philippines, 186
online education, 69, 94–95, 102, 113,
149, 153, 164, 170
online learning environment (OLE)
facility, at OUHK, 47, 54
online orientation course, at GSIS,
150–51
open and distance learning (ODL),
xvii–xviii, 169
OUSL contribution to national QA
framework development, 253
quality framework for, 248–50
Open Education Resource (OER)
movement, xviii
Open Entry Admission System
(OEAS), at WOU, 202

open universities (OUs)
origin of, xvii
Open University Malaysia (OUM), 200
approval for open entry system, 261
establishment of, 260
internal QA measures to address
delivery issues, case studies
assignment practices, 269–72
learning modules supply, 265–69
ODL programs at, 261
quality management at, 261–65
future challenges, 273–74
lessons learned, 272–73
Senate responsibility, for academic
matters, 260
starting of operations in 2001, 260
Open University of China (OUC),
155, 277. *See also*, China Central
Radio; Radio and TV University
(RTVU) system; TV University
future challenges, 166–67
lessons learned, 165–66
quality assurance framework
issues and challenges in, 164–65
learning support services, 160–61
teaching and learning
environment, 163–64
teaching management, 162–63
teaching process management,
158–60
teaching resources development
and management, 156–58
Open University of China system
(OUCS), 94
Open University of Hong Kong
(OUHK)
assessment on course-by-course
basis at, 50–51
course development and approval,
47–48
course review and revision, 48–49
establishment of, 42
lessons learned, 55–56
new developments impact on
collaborative programs, 54
e-learning introduction, 53–54
face-to-face provision
introduction, 51–53
programs in Mainland China, 53

philosophy on quality, 42–43
program development and approval
at, 45–46
program review, revision, and
revalidation, 46–47
quality assurance at, 44–45
quality concerns in, 43–44
student support, 49–50
Open University of Japan (OUJ), 141
Open University of Sri Lanka
(OUSL)
central campus at Nawala, 243
Distance Education Modernization
Project (DEMP), 243–44
establishment of, 242
issues and problems faced and
solutions, 252–54
learner profile, 243
lessons learned, 254–55
national QA framework and
performance indicators,
development of
experience in QA toolkit use,
250–51
framework for ODL, 248–50
performance indicators, 250
organizational structure of, 243
quality assurance and enhancement
mechanism at
interventions in distance learning
and teaching, 246–48
management structure, 244–46
Open University, UK (OUUK), xvii,
xxi, 44
outline program proposal (OPP)
document, at OUHK, 45

Pakistan
Higher Education Commission
(HEC) establishment in 2002,
220
higher education rate in, 220
universities chartered by
government, 221
virtual university of, 222–23
Pakistan Bar Council (PBC), 221
Pakistan Council of Architects and
Town Planners (PCATP), 221

Pakistan Engineering Council (PEC), 221
Pakistan Medical & Dental Council (PMDC), 221
Pakistan Nursing Council (PNC), 221
Pakistan Pharmacy Council (PCP), 221
Pakistan Veterinary Medical Council (PVMC), 221
Philippine Accrediting Association of Schools, Colleges and Universities (PAASCU), 183–84
Philippine Association of Colleges and Universities' Commission on Accreditation (PACUCOA), 183
Philippine Women's University, 182
Plan–Do–Check–Act (PDCA) process, at STOU, 34
planning, monitoring, and evaluation system (PriME), 112
Polytechnic University, Philippines, 182
private education institutes (PEIs), 3–4, 23
Private Higher Educational Institutions Act 1996, 260
private sector universities, in Singapore, 3
program approval process, at IGNOU, 174
program definitive document (PDD), 11–12
program delivery, of UniSIM, 12–13
program development and implementation stage, of AE quality assurance tools during, 123–26
program development process at OUHK, 45–46 at UNiSIM, 11–12
program planning and development quality assurance intervention during, 208–9
Program Review and Validation Committee (PRVC), 46

QA Toolkit, use by OUSL, 250–51
Quality as Process (QaP) management system, 112

Quality Assurance and Accreditation Council of the UGC (QAAC-UGC), Sri Lanka, 241–42, 245
Quality Assurance and Accreditation System (QAAS), Japan, 139 objectives of, 140 results by agency, 141
Quality Assurance Committee (QAC), 42
Quality Assurance Framework for Universities (QAFU), Singapore, 3, 8
quality assurance (QA) system, xix, xviii, 275–77
at AutoEver (AE), Hyundai e-learning programs and course evaluation systems, 120–23 tools during development and implementation stage, 123–26
in cyber universities of Korea, 60–65
definition of, xvii
in higher education, 81
development by UGC and CVCD, Sri Lanka, 241
at Graduate School of Instructional Systems (GSIS), Japan, 143–51
at Indira Gandhi National Open University (IGNOU), India, 172–77
at Open University Malaysia (OUM), 261–65
at Open University of China (OUC), 156–64
at Open University of Hong Kong (OUHK), 44–45
for content development, 69–71
for faculty support, 73
for implementation of course, 70–72
for integrity in academic system, 73–74
for interaction, 73
for mentoring program, 74
for tutor system, 72
at Open University of Sri Lanka (OUSL), 244–51

at School of Distance Learning for
Medical Education (SDLME),
China, 96–100
at Sukhothai Thammathirat Open
University (STOU), Thailand,
30–31
evaluation results, 35–36
integration of internal and
external QA, 33–35
internal QA, 31–33
issues and concerns in internal
and external QA systems, 36–37
lessons learned for effective
development of, 38
total quality management (TQM)
model integration in, 39–40
at Universitas Terbuka (UT),
Indonesia, 84–91
at University of the Philippines-
Open University (UPOU),
185–92
at Wawasan Open University
(WOU), Malaysia, 205–14
quality culture, promotion of, 281–82
Quality Enhancement Cells (QECs),
Pakistan, 221–22
quality enhancement (QE), 200
quality interventions, at OUSL
in distance learning and teaching,
246–48

Radio and TV University (RTVU)
system, 155–60, 162–63. See
also, Open University of China
(OUC)
recruitment policy and procedures, at
UNiSIM, 13
resourcing, 20–21

School of Distance Learning for
Medical Education (SDLME),
Peking University's China
achievements of, 101
challenges and lessons learned,
101–3
establishment of, 95
goal of, 95
partners in, 95

programs in, 96
quality assurance
activities of, 98–100
mechanisms and procedures at,
95–96
structure of, 97–98
self instructional material (SIM),
172–73
Shinshu University's Graduate School
of Science and Technology,
Japan, 141–42
Singapore Institute of Management
(UniSIM), 3. See also, Council
for Private Education (CPE);
Quality Assurance Framework
for Universities (QAFU),
Singapore
academic QA framework, 9–11
academic standards of, 8
assessment practice adopted by,
14–15
Associate Faculty (AF) members
in, 13
benchmarking with other
universities, 17
e-learning establishment at, 13–14
establishment of, 4
future challenges before, 22–23
general issues faced by, 17–18
internal academic audit process, 8
organizational structure of, 5–8
partnerships in programs, 15
program and course development,
11–12
program delivery, 12–13
quality assurance issues faced by,
18–22
quality management at, 8–11
"second chance" philosophy, 4–5
student feedback on education
system, 16–17
supporting infrastructure, 15–16
undergraduate programs for
working adults and adult
learners, 4
values of, 5
Singapore Institute of Technology, 3
Singapore Management University, 3

Singapore tertiary education, 3
Singapore University of Technology
and Design, 3
Soros Foundation, 105
staff development opportunities, at
GSIS, 145–47
staffing, 20
Staff Training and Research Institute
of Distance Education
(STRIDE), 173
Standards and Industrial Research
Institute of Malaysia (SIRIM),
262
Standards for Establishing University
(SEU), Japan, 139
state open universities (SOUs), 169–70
story-centered curriculum (SCC),
GSIS, 149–50
student distribution, in cyber
universities of Korea
by academic background, 59
by age, 59
student support system, at HYCU,
68–69
Sukhothai Thammathirat Open
University (STOU), Thailand,
276. *See also*, Malcolm Baldrige
National Quality Award
model
addressing of individual and
societal needs, 30
Asian Management Award in 1996
to, 30
establishment of, xix, 29–30
internal and external QA measures,
relationship between, 29
quality assurance (QA) system at,
30–40

teaching and learning environment, at
OUC, 163–64
Teaching Excellence Awards, 13
teaching management, at OUC,
162–63
teaching process management, at
OUC, 158–60
teaching resources development and
management, at OUC, 156–58

technology-enhanced open distance
learning (TEODL) mode, 202
ThinkQuest, in Mongolia, 106
Tokyo University of Career
Development, 142
total quality management (TQM)
model
integration in STOU QA system, 39
tuition fees, 5, 75, 183, 227
tutor-marked assignments (TMAs),
12
tutor system
element of OUHK learning support
system, 49–50
QA at OUHK, 72
tutor training and management
system, at WOU, 211
TV University, 155. *See also*, Open
University of China (OUC)

undergraduate programs, at HYCU's,
66
unexpected demands, E-learning
evaluation process for, 129
UNICEF, 105, 284
Universitas Terbuka (UT), Indonesia
challenges and lessons learned
balancing standardized ISO
procedures and human
working culture, 91–92
building personnel and internal
audit capacity, 92
need for ensuring quality, of
academic content, 91
conducting online examination from
2006, 83
establishment of, 82
faculties in, 82
learning medium in, 83
open registration and learning
system, implementation of, 83
quality assurance at
certification with ISO 9001
Standards, 87–91
policies of, 84–87
University Academic Management
Committee (UAMC), Malaysia,
260, 263, 267–68, 270

University Grants Commission
  India, 169
  Sri Lanka, 241
University Grants Committee (UGC),
  42
University of Digital Content, Japan,
  142
University of South Africa, xvii
University of the Air. *See* Open
  University of Japan (OUJ)
University of the Philippines–Open
  University (UPOU), 182–83
  lessons learned, 192–94
  quality assurance at
    in course delivery. 188–89
    in course development, 187–88
    issues and possible solutions,
      189–92
    in program proposal
      development, 185–87

video conferencing centers, in
  Mongolia, 107
virtual university of Pakistan (VUP),
  222–23, 278
  quality assurance and enhancement
    measures at
    lessons learned, 236–37
    procedures specified as rules,
      226–28

QA/QE system leveraged by
  technology, 224–26
student assessment system, case
  study, 228–32
Visayas State College of Agriculture,
  Philippines, 182

Wawasan Open University (WOU),
  Malaysia, 200, 278
  academic programs in, 203
  course delivery model of, 204–5
  delivery of courses through TEODL
    mode, 202
  education model, 202–3
  establishment of, 202
  first student intake in 2007, 202
  multi-entry and multi-exit
    progression pathways, 203
  quality assurance at
    issues and solutions, 214–17
    lessons learned, 217–18
    management, 206–7
    policy, 207–8
    practice of, 208–12
Wireless Application Protocol (WAP)
  service, library for HYCU
    students, 69
World Bank, 106
World Wide Web, xvii, xxi